Chapter One

INTRODUCTION

I DISTINCTLY REMEMBER in grade school how easy mathematics was for me, and yet how challenging it was for many of the other children in the same classroom. The same scenario would also appear both in my junior high years, and then off to my high school years where I distinctly remember not even having to study math. Studying at different universities left me with the same impression of why some students found math easy and most found it increasingly challenging. As I went on to graduate school and studied higher mathematics, I then began to see a pattern in myself. As the math became increasingly theoretical in graduate school, the less I became aware of its implications and applications. I then realized how difficult it was to learn math at a high level that was solely based on theory. This was the theme played out before my eyes as I worked my way through graduate school to get a graduate level degree in mathematics.

It was clear that there was something missing in math curriculum. I could see that when math curriculum was theoretical and general in nature, students had a difficult time learning the math concepts. When math was more applied and structured to real life, students became more interested in math and had a higher interest in solving the math problems. I particularly remember finishing graduate school, thinking I really needed to be in an industry that uses math, so I could see how math works in real life. After graduate school, I then went on to work for a large insurance company where I worked in their management information systems department just north of San Francisco, California. Having worked for this insurance company for 15 years, I saw how math was used in a business and application setting. During my tenure in the insurance industry, I started teaching math at a junior college in central California. It was then that I realized I needed to shift my focus and begin a journey that would lead me to teach and research contextualized mathematics at the college and university levels.

At that point in my career, I could see that there was certainly a need to begin to reflect and think about how math could be applied and used in a new type of math curriculum. I was fortunate enough to do research at a college in Kirkland, Washington i.e., Lake Washington Institute of Technology. My work centered around the research and development of contextualized math curriculum that was applied within technical fields. The key for my research centered on developing math that students could apply in their technical fields. I then went on to research other applied math curricula at other colleges in the state of Washington. My research, my teaching experience, and my love of teaching and learning mathematics form the impetus for this book.

There is a definite need for a new type of math curriculum. From a broader perspective, students are entering colleges and universities with a skill deficiency in mathematics which affects their ability to complete their college math requirements successfully. College students are struggling to complete required college mathematics standards after they enter college (Albritton, Gallard, & Morgan, 2010; Bonham & Boylan, 2011; Hern, 2012).

The reason students need to complete their college and math courses is that getting their degree requires them to successfully complete these specific courses. Furthermore, students who have previously taken face-to-face math classes are now enrolling in online math courses with the use of new technologies, which investigators should consider when researching student completion rates (Ashby, Sadera, & McNary, 2011; Jones & Long, 2013). According to research by the Community College Research Center of online education within the state of Washington and the state of Virginia, the completion rate for online remedial math was 19% lower than completion rates in face-to-face classes (Crawford & Persaud, 2013). Furthermore, in 2008, three-fourths of students who started remedial math in college never finished their remedial math successfully (George, 2010).

One of the challenges college students face is the linking of math concepts, organizing mathematical ideas, and solving math problems to complete their college remedial math and college math courses successfully (Albritton et al., 2010; Bonham & Boylan, 2011; Diaz, 2010). If educators fail to address this student math problem through new curricula such as college contextualized math curriculum, this will likely lead to high failure rates in college and university math classes, which affects students' ability to complete their certificate or degree requirements for graduation (Howard & Whitaker, 2011; Sheldon & Durdella, 2010).

In my research, I focused on developing contextualized mathematics in math curriculum. Therefore, it is first important to understand the concept of contextualization. Contextualization is a method used for math curriculum design within lesson plans and as a pedagogical practice in face-to-face math classes. Many researchers are studying the effects of using contextualization in the teaching and the learning process for mathematics learning (Bottge & Cho, 2013; Perin, 2011; Young, Hodge, Edwards, & Leising, 2012).

Contextualized math refers to the teaching of mathematics problems that emphasize real-life situations. For example, an instructor could develop a contextualized math assignment on exponential functions and equations by having students work on an Internet-based problem to determine the monthly payment for a new car at a given price and interest rate for a specific loan duration. In this example, instructors would teach students how to compute with exponential formulas to come up with the monthly amount.

Contextualization is defined as the development of math problems that are authentic and related to real world applications or math problems that are connected to their future careers (Bottge & Cho, 2013; Bottge, Ma, Gassaway, Butler, & Toland, 2014; Valenzuela, 2012, 2014). Furthermore, students often have difficulty solving computational math problems solely focused on formulaic computations, or without any real-world connections (Khiat, 2010; Puri, Cornick, & Guy, 2014). From my experience in teaching college and university mathematics, many students struggle in understanding and learning math concepts. In fact, for many students, their problem-solving and computational skills are not adequate for them to succeed in calculating mathematical computations (Bottge, Grant, Rueda, & Stephens, 2010).

Contextualized mathematics provides a foundation for researchers, instructors, and administrators to implement and study how students learn mathematics in the classroom. Moreover, the use of college contextualized mathematics curriculum is of importance to students, educators, and researchers because of the high failure rates in college math and college remedial math education (George, 2010; Howell, 2011). Researchers found there to be only a 25% success rate in remedial college math education (George, 2010). Furthermore, Howell (2011) indicated that the high failure rates in college math cost four- year colleges and universities between $435 and $543 million annually. Other researchers such as Bonham and Boylan (2011) noted that college students had to take an average of 10 credit hours of remedial college math before starting their college level mathematics courses.

The use of contextualized mathematics helps students learn mathematics. For instance, Asera (2011) found the use of new college math curriculum with contextualized math components and accelerated learning as another avenue for addressing the problems of non-completion. Furthermore, Burrows, Wickizer, Meyer, and Borowczak (2013) investigated the interest in math by using contextualized applied problems, and they found students improved their content area learning after exposure to the contextualized math curriculum. Similarly, researchers such as Mhakure and Mokoena (2011), determined that with a such a curriculum, students were able to relate to the math problems in a real life setting.

Prior research has shown that a contextualized math curriculum is an effective form of teaching mathematics (Bellamy & Mativo, 2010; Bottge & Cho, 2013; Deed, Pridham, Prain, & Graham, 2012). Therefore, because of the success of such mathematics, this math curriculum is an area of further investigation for college math curricula research (Asera, 2011; Merseth, 2011; Perin, 2011). College administrators, instructors, and researchers are very interested in using new math curriculum to address the high failure rates in college math because of the costs to both colleges and students (Howell, 2011). From a theoretical and research perspective, the use of a college contextualized math provides an avenue to further investigate how students construct mathematical knowledge.

SUMMARY CONCEPTS:

- Math concepts and math lesson plans based on theory is challenging for students to learn and understand

- College students take an average of 10 credit hours of remedial college math before they enroll in their required college level mathematics courses

- The use of contextualized mathematics helps students learn mathematics

- Contextualized math curriculum is an effective form of teaching mathematics

Chapter Two

Constructing Math Knowledge

Within a math class, students construct their mathematical knowledge through the various assignments, questions, and activities posed by the math instructor (Fast & Hankes, 2010). The use of linking concepts via the integration of concepts in the learning process arose from constructivist learning theory (Dewey, 2011; Piaget, 2001; Vygotsky, 1978). An example of the use of constructivism in math education is an instructor's use of anchored examples, which are connected to prior concepts and are further developed by the instructor (Fast & Hankes, 2010). For instance, a math instructor anchors the solution steps to solving a monthly payment amount for a given interest rate loan with the solving of complex fractions. Furthermore, the concept of constructivism is based on helping students organize, categorize, and cognitively think about topics (Dewey, 2011; Piaget, 2001; Vygotsky, 1978). A math instructor can effectively use constructivism via the mathematics lesson plan when discussing the introductory concepts to advanced methods of mathematical modeling, where students learned at different levels from an introductory to an advanced level (Milner, Templin, & Czerniak, 2011).

How students construct knowledge is the key to the development of contextualized math curriculum. The reconstruction of knowledge is an essential concept in constructivism. The reconstruction of knowledge and organization of knowledge are part of constructivism, and a student builds their knowledge and understanding through constructivism (Al-Huneidi & Schreurs, 2012; Piaget, 2001; Vygotsky, 1978). Students are reconstructing new knowledge for themselves through the development of knowledge based on students' experiences (Al-Huneidi & Schreurs, 2012; Piaget, 2001; Vygotsky, 1978). Furthermore, within math education, math instructors develop math material from mathematical definitions, axioms, and theories. The formal axioms, theory, and definitions are often overwhelming for students, and hence, the instructor's role is to facilitate the students' reconstruction of information into a manageable and understandable form within the curriculum. This seems to be where the difficulty lies. *When the focus of math becomes the definitions, axioms, and theoretical nature of math, students become less interested and can be lost in the application of the math concepts.*

Whether in an online or face-to-face math course, instructors are asked to find new ways of constructing knowledge for their students. For instance, students can write down mathematical information such as definitions and formulas to further their studies and work on problems. Researchers indicate that students build knowledge based on what they have constructed and what they already know (Ulrich, Tailem, Hackenberg, & Norton, 2014). Therefore, when using contextualized math (i.e., applied math problems), this keeps students from using algorithmic learning of formulas and definitions because the emphasis is on the application and building of new information rather than memorization of it.

The theory of constructivism provides a basis for further examining why a contextualized math curriculum helps students learn math. For instance, the use of modeling a savings account with coins, which is a form of contextualized mathematics, allows students to add and subtract using mental models of signed numbers via a constructivist mathematics curriculum (Ulrich et al., 2014). Contextualized mathematics curriculum includes the development of mathematical problems in context to a real problem. Investigators see this same frame of reference in a constructivist research study in which one recommendation is to develop the actual mathematical problems in context to enable students the ability to learn through the students' interaction with the problem (Sahin, 2010). In addition, the use of diagrams, graphs, and visual aspects of displaying mathematics provides a broader context and conceptual framework for students in a constructivist learning methodology (Krummheuer, 2013). This type of manipulative and visual use of math is certainly evident when students work on a mathematical problem by developing a model, working with the data, and students working together as a team to come up with a solution (Fatade, Arigbabu, Magahi, & Awofala, 2014). In this respect, constructivism is a framework for learning math through an active learning process (Grady, Watkins, & Montalvo, 2012).

From a learning perspective, the constructivist theory explains how students learn math through a contextualized math curriculum. One of the challenges with students constructing and developing their knowledge in math is the different methods students used in solving problems, which inhibits effective problem-solving taught by instructors (Ulrich et al., 2014). When the math instructor provides a problem in context to a real life problem, which is contextualized mathematics, students explore and benefit from this style of learning (Perin, 2011). Through this sort of focus on collaboration and working together, students explore a specific topic and problem.

Similarly, math instructors enhance the learning process in mathematical context or contextualized mathematics through the active evaluation, discourse, and probing of the problem among students (Sahin, 2010). Many of the real-life problems are too complex for students and individuals to work on independently. Hence, a collaborative approach is most effective in problem-solving mathematical problems as students construct mathematical knowledge when they work together (O'Shea & Leavy, 2013). In a similar fashion to the implementation of constructivist methodology, Wood (2010) sees *teaching students in contextualized math along with the teaching of professional skills, teamwork, and communication as effective teaching techniques to prepare students for the workforce.*

A contextualized math curriculum gives students an opportunity to challenge their current beliefs about math by seeing the connection of real life math to their own lives. Students that see a conflict from their prior beliefs in mathematics see conceptual learning happen only when confronted with new information (Hennessey, Higley, & Chesnut, 2012).

In another instance, students could be asked to calculate the loan payment for the purchase of a home or a new car given the interest rate and loan duration using an amortization schedule in Excel. This new information could provide the student with knowledge of how much the loan would be, the payment, and if a home was affordable in the future. As described in this example, students would construct their own knowledge and ask questions about the amortization process. Zain, Rasidi, and Abidin (2012) conclude that students work best when allowed to work together on mathematical problems, while asking and answering questions among themselves.

By using a contextualized math curriculum, students are able to construct mathematical knowledge. Further, understanding the role of the theory of constructivism by implementing a contextualized math curriculum, educators, researchers, and administrators will be able to develop effective ways for students to learn and pass their math course requirements.

As you develop your understanding and knowledge of mathematics curriculum, there are key terms to focus on that are listed below. These terms will further your knowledge regarding how mathematical learning occurs for students whether in an online or face-to-face classroom.

Key Terms

Algorithmic Learning. Algorithmic learning is a method of teaching mathematics that emphasizes the teaching of formulas and memorization as part of the math curriculum as opposed to teaching a conceptual form of math in which students can generalize their findings to larger math concepts (Battey, 2013; Fan & Bokhove, 2014).

Anchored Math Instruction. This is a form of teaching mathematics in which the math content is anchored to a real-life problem situation. For instance, students may be asked to build a skateboard ramp using materials, measurement, and the Pythagorean Theorem of a right triangle (Bottge et al., 2010).

Authentic Learning. Authentic learning is based on connecting a problem or situation in life that a student can relate to when completing their assignment inside or outside of class (Dennis & O'Hair, 2010).

Contextualized Mathematics and Learning. In the context of this book, a contextualized math curriculum is one in which the mathematics curriculum includes math problems that are based on real life situations and real life problems that students can relate to their own daily lives or their future careers. Contextualized learning is known by other terms such as embedded instruction, integrative curriculum, problem-based learning, anchored instruction, curriculum integration, modeling-based math, and work-based learning. Context and contextualization in this framework are in the form of relating mathematics to students' lives, work, community, and school (Perin, 2011).

Developmental Education and Math Remedial Education. Developmental education and remedial math education refer to pre-college courses typically covering material the student learned in K-12. These courses are taken by college students to fulfill a requirement before entering college level coursework (Asera, 2011).

Math Conceptualization. In this book, math conceptualization will refer to the ability of the student to draw connections to the mathematical material learned. Furthermore, students' math conceptualization goes beyond the memorization of mathematical formulas. In essence, math conceptualization is not solely knowing the mechanical nature of math, but also knowing how to apply the formulas and problems in different types of applications (Khiat, 2010).

Math Literacy. Math literacy is another form of student learning called numeracy, which is the ability of a student to take mathematics, data, and analysis, and then apply it to problem-solving using critical thinking (Mhakure & Mokoena, 2011).

Math Self-Efficacy. Math self-efficacy is a measure that identifies the level of confidence a student has in their ability to solve mathematical problems. Students with high levels of math self-efficacy have improved on their math achievement scores (Briley, 2012).

Remediation. Remediation refers to the identification of specific math skills or writing skills of a college student and having the student address those skill deficiencies through instruction at a college. Remediation is often referred to as developmental education in college (Bahr, 2012).

Summary Concepts:

- Students construct their mathematical knowledge through the various assignments, questions, and activities posed by the math instructor

- An effective technique is for an instructor to use anchored examples, which are connected to prior concepts and further developed by examples and context by the math instructor

- Theory, axioms, and definitions are often overwhelming for math students, and hence, the instructor's role is to facilitate the students' math understanding into small and manageable units of math learning

- When the math instructor provides a problem in context to a real life problem, which is contextualized mathematics, students explore and benefit from this style of learning

- A contextualized math curriculum gives students an opportunity to challenge their current beliefs about math by seeing the connection of real life math to their own lives

Chapter Three

Research on Learning Mathematics

The learning of mathematics through student math performance is an avenue to determine how students learn math. To better understand the learning of mathematics, researchers have used different types of studies to analyze student math learning. Baird (2011), a researcher, used a quantitative research method to analyze data in the state of Washington in order to identify student performance in math. The researcher captured the data as part of the analysis of a standards-based reform, which was part of No Child Left Behind. Schwols and Miller (2012) investigated mathematical content related to the current integration of Common Core State Standards in math curriculum. Researchers found connections between the students' tenth grade math achievement and their early math achievement in the seventh grade. From a different perspective, another researcher, Khiat (2010), noted students learned math through their association and conceptualization of mathematics via procedural methods.

Another avenue for learning mathematics is through cooperative learning. Capar and Tarim (2015) investigated cooperative learning for students learning mathematics. Cooperative learning is defined as a method of studying by students where they interact, discuss, and work together for math problem-solving. The researchers employed a meta-analysis research technique that integrated quantitative analysis in the research method. Similarly, Deed et al. (2012) investigated the learning of mathematics in group work activities. The difference is that Deed et al. also incorporated the use of math through real life contextualized applications. Capar and Tarim collected experimental research data from 1988 to 2010. Capar and Tarim integrated 26 research studies in their investigation ranging from pre-school learners to university students. The investigators compared the cooperative learning environment versus learning in a classroom setting in a traditional form of learning math without any cooperative learning. Capar and Tarim revealed that the use of cooperative learning helped students in math achievement in comparison to the traditional form of learning mathematics. Likewise, Deed et al. discovered that group learning engaged the students in learning math through group activities.

Math instructors incorporate the use of videos to assist students in learning mathematics. Kinnari-Korpela (2015), a researcher, investigated the use of videos to improve the learning of mathematics by students. One of the central issues identified by the researchers was the lack of math skills in students entering a university. The investigator identified algebra skills as chief skills lacking by the entering students, which the researchers felt could require a connected teaching methodology.

A concern among math instructors is the time spent in class teaching remedial mathematics to students rather than the math content required. Redmond et al. (2011) researched the use of engineering topics to assist students in learning mathematics. Kinnari-Korpela incorporated 18 math videos as an introduction of math topics and as a component in lesson plans. Kinnari-Korpela indicated that 89% of the students felt that learning math from the videos was important to them. The investigator noted that some of the course material could be replaced by the videos because of their effectiveness. The researcher also indicated that 65% of the students agreed that the videos helped improve their motivation for learning math.

The use of manipulatives can also assist students in the learning of mathematics. Chun-Yi and Ming-Jang (2015) studied the use of examples in mathematics with manipulatives to assist students in learning mathematics. The understanding of fractions and equivalent fractions by grade school students was the purpose of the investigation. Manipulatives provided students a physical representation of the math concept. For instance, a circular object could be cut into four equivalent pieces to describe fractions. Also, there were working examples of math problems with solution steps, which provided a step-by-step instructional process for students to follow. Virtual Manipulatives are physical representations using computers or online programming to display math concepts virtually. In contrast, Wake (2014) researched the use of modeling and technology to demonstrate and integrate the use of mathematics for students. In addition, Chun-Yi and Ming-Jang applied a quasi-experimental method in the study. There were 100 students from the fifth grade in Taipei City. Chun-Yi and Ming-Jang identified that the use of virtual manipulatives and worked examples aided students in understanding the concepts of fractions. In a similar fashion, Wake concluded that allowing students to explore math through modeling and spreadsheets aided them in learning mathematics.

Math instructors accomplish the teaching of mathematics through effective learning activities in the classroom. For instance, Hasan and Fraser (2014) researched the use of activity strategies in mathematics classrooms at the college level for effectiveness. They emphasized a change in roles for the students in the classroom. Instead of a passive role, students were encouraged to ask questions and engage the math topics in class. The researchers were particularly interested in students who had difficulties in learning mathematics at a young age. The study was conducted in the United Arab Emirates at a technical college, the Higher Colleges of Technology. Hasan and Fraser collected data through interviews and observations. The activity strategies included games and hands-on work with the math students. They revealed that the use of activity strategies in the classroom motivated the students and changed the students' views of learning mathematics.

Math instructors' use of cooperative learning through peer-led teams is another strategy for teaching and learning mathematics. Reisel, Jablonski, Munson and Hosseini (2014) studied the use of Peer-led Team Learning (PLTL) for undergraduate engineer students taking college mathematics. They emphasized group work—that is, students working together on mathematical problem-solving in a team environment. Similarly, Trinter, Moon, and Brighton (2015) researched problem-based learning, which incorporated the use of cooperative learning to motivate students in the math class. For Reisel et al., they conducted their study for incoming freshman students attending the University of Wisconsin at Milwaukee. One of the central reasons for the research at the research site was the low graduation rate, which was 26.3 percent of computer science and engineering students at the university. The principal reason for the low graduation rate was the inability to complete the college math requirements. In addition, the researchers figured that 50 to 70 percent of the incoming freshman in computer science and engineering started at the remedial and pre-calculus levels.

The students worked on mathematics problems as a group with an upper-class engineering student as the facilitator of the group. Training was provided to the group facilitators. The overall goal was to provide a cooperative environment for the math students to work together and gain confidence in solving math problems. The researchers noted that students with increased involvement in the groups also improved their grades in math. The group settings yielded the most benefit to students in calculus classes.

The conceptual learning of mathematics provides a foundation for students. Sidney and Alibali (2015) investigated the use of conceptual learning and prior-knowledge comprehension in mathematics. The purpose of the study was to understand how prior knowledge affects the learning of new mathematical concepts. The researchers considered the use of structural understanding as a way of transferring mathematical understanding to new mathematical problems. In addition, the researchers distinguished between procedural and conceptual knowledge. Procedural knowledge involves specific steps to solve a problem whereas conceptual understanding included the application of mathematical knowledge to new situations or problems.

In Sidney and Alibali's research, the specific area of transfer of knowledge was with the division by fractions in mathematical problems. One of the challenges with these types of problems was that students used prior knowledge, which did not translate to the division by fractional amounts based on prior fractional learning. There were 100 students at the fifth and sixth-grade levels, drawn from the U.S. Midwestern region. The researchers utilized pretests and posttests to identify the knowledge of students in the division by fractional quantities.

Two conceptual problems were also included in the tests to indicate the level of conceptual application in math story problems. Investigators provided students an oral lesson on division, and they were given math practice-problem worksheets. The researchers used a quantitative method to analyze the pretest and posttest data. Sidney and Alibali determined that with the math problem worksheets and lessons provided, a positive reinforcement of conceptual knowledge and understanding to the students enabled them to transfer their mathematical knowledge to the division-fraction problems. Researchers indicated that students enjoyed the conceptual learning and application of mathematical concepts to real life situations.

The use of technology in a math classroom can aid in the learning of mathematics. Erbas, Ince, and Kaya (2014) investigated the use of interactive white boards and graphing software to aid in understanding of quadratic functions and graphs. The researchers included 65 high school graduates from Turkey preparing for university entrance exams. The students' ages ranged from 18 to 20 years old, and students were selected based on the high school grades and prior years' entrance exams. The researchers developed a graphing test to measure the understanding and graphing capabilities of the students with quadratic functions.

The graphing software, NuCalc, provided the students the ability to graph by entering an equation into a computer application. Similarly, other math researchers used computer-assisted software to assist students in solving and working through math word problems. Moreover, Erbas et al. used interactive whiteboards to convey lecture material and graphs. *Furthermore, the researchers revealed that the use of the technology and software improved the attitudes of students towards learning mathematics.* The researchers concluded that the students' understanding of graphical representations of quadratics had improved because of the use of the graphing software. Similarly, other math researchers have concluded that the use of technology assisted remedial math students in improving their math skills.

Researchers and instructors are incorporating new teaching models in the math classroom to improve the effectiveness of learning. Clark

(2015) investigated the use of a flipped model in a mathematics classroom to improve the engagement and mathematical learning for secondary math students. The flipped model of teaching included the use of technology for lectures outside of the classroom and math assignments and activities as a focus within the class. This is in contrast to the traditional approach of teaching math in which the instructor covers the math lecture during the classtime and then assigns homework for students to complete outside the classroom. Likewise, Wake (2014) noted that instructors should use technology such as spreadsheets to assist students in learning mathematics. In Clark's study, two algebra high school classes formed the sample in the study. Clark used the flipped model as the intervention to determine its effect on students' engagement and learning of mathematics. The intervention lasted seven weeks during the two algebra courses. The in-class instruction included real-life applications of mathematics. At the end of the seven weeks, a unit exam was given to both classrooms. Furthermore, twelve students were randomly selected for interviews. The interview questions were also field-tested. Students who did not have Internet access outside of the classroom were able to view DVDs and other course material through flash drives. Students felt that with the flipped model of teaching, the quality of math instruction improved. Students also enjoyed the use of cooperative learning in the flipped model classroom. Clark showed students more actively engaged in the flipped model classroom than in the traditional math class structure.

The use of formative assessments and self-regulated learning can have a positive effect on students learning math. Hudesman et al. (2014) studied the use of formative assessments along with self-regulated learning to improve the process of mathematics for college math students. The formative assessments in the study were math quizzes. Similarly, Nijlen and Janssen (2015) incorporated assessments in their research study on learning mathematics. However, Nijlen and Janssen used both non-contextualized and contextualized mathematics assessments in their research study, whereas Hudesman et al. provided no differentiation of contextualization in their research analysis. Within Hudesman's et al. research, the self-regulated learning aspects of the formative assessment program involved students completing a self-reflection form, which provided them an opportunity to review the instructor feedback on their quiz and to gain further mastery of the math.

The instructors used the self-reflection form as a way to develop further activities and assignments for the students. Hudesman et al. described an Enhanced Formative Assessment and Self-Regulated Learning (EFA-SRL) multi-campus program to improve community college students' achievement in developmental mathematics courses. Moreover, the program students earned higher mean grades and achieved higher pass rates than their counterparts in baseline classes. The investigators selected two of the community colleges located in the Northeast and Midwest for the research study. There were a total of seven courses in developmental mathematics at the two-year colleges. The researchers found that the use of formative assessment and a self-regulated learning math program improved the achievement scores of students in developmental math. Furthermore, in comparison with a baseline developmental math class, the students had a higher mean grade than students did in the baseline course. In contrast, Nijlen and Janssen's research showed that added engagement was evident in assessments, which included contextualized real-life applications of mathematics compared to the non-contextualized mathematics assessments.

By researching the experiences of students learning math, researchers and educators will be in a better position to understand how students best learn math. Roykenes (2015) used narrative qualitative research to study how 11 nursing students learned mathematics. The goal of the research was to study the prior experiences of the nursing students that influenced the current understanding and knowledge of math. Roykenes used narratives developed by the nursing students as a basis for the data collection. Norwegian university nursing students were selected because of the importance of mathematics in nursing calculations and college nursing exams. Nursing students wrote the narrative stories during their initial semester in the nursing program. The researcher incorporated a thematic method to analyze the narratives for emerging concepts and ideas.

Roykenes found that most students had a positive experience with learning math at the primary school level. However, Roykenes also concluded that the students' perceptions of math were affected by the class size, noise in the classroom, and the difficulty of the math topics as the students entered secondary school. Providing homework assistance to students in class made a difference in their learning of mathematics and ensuring that the math was relevant to the students' lives helped in the learning process. The researcher recommended using mastery learning by focusing on smaller learning tasks, using smaller cooperative learning groups, and using an interactive approach to learning math by practicing the math calculations.

Providing students a feedback process can enable students to use the feedback in the learning processes of mathematics. Duhon, House, Hastings, Poncy, and Solomon (2015) researched the use of accuracy feedback as an intervention method for the improvement of math learning for students. A Midwestern grade school was the site for the study, and there were 48-second grade students in the research project. The intervention method was a computer program in the school's computer lab. The computer program was set up with different options that provided both no feedback and immediate feedback modes.

In contrast, Spradlin and Ackerman (2010) used computer-assisted learning modules in an intermediate college algebra class in comparison to teaching the same class without computers. For Duhon et al., the feedback to the students was in the form of accuracy feedback on the completion of the math problem. The students worked on addition and subtraction problems with the researchers tracking the number of correct problems answered. The researchers found that the immediate feedback and intervention accuracy feedback improved the math understanding of the students exposed to those interventions.

Math educators and researchers are investigating how students' interaction with each other can improve the construction of math knowledge. For instance, Hanim, Mohd, and Zainol (2012) investigated the use of Student-Centered Learning (SCL) as a method of learning mathematics with the constructivist theory as the framework of the study. SCL was defined as a learning method with students as the initiators in the learning process. Constructivism was the foundational theory of the study where the students incorporated problem solving and generating solutions and ideas about the math studied within the class.

Within constructivism, the student is the initiator of knowledge, and the student is building knowledge based on their prior learning. Another term for SCL was cooperative learning. Students were able to learn from each other as they questioned and discussed the various solutions of the math problems on which they worked. Instructors in the study believed that collaborating with pharmacy and medical fields would help relate the math within those fields for students. The researchers recommended that instructors consider how students think about the math of each lesson plan. Allowing students to construct knowledge through a hands-on approach of learning mathematics was another recommendation by the researchers. Researchers showed that the use of the cooperative learning in a math class aided students in their improvement of math learning and achievement.

Gaming platforms are now becoming a foundation for learning math in the classroom. Huang, Huang, and Wu (2014) studied Digital Game-Based Learning (DGBL) as a method for learning mathematics in the classroom. Investigators incorporated a computer tablet with a diagnostic into the study. Specifically, the researchers were interested in studying math, anxiety, and motivation when using a digital strategy for math students. Researchers noted the system provided feedback for the students as they worked through the math problems. The math transactions involved the use of subtraction and addition for grade school students learning mathematics.

The diagnostics feedback mechanism in the system provided students the ability to learn from their math errors. The students revealed in the interviews that the system reduced their math anxiety. The practical application of math by using it to purchase items in a store also motivated students to learn math.

Math educators are studying math problem-solving to aid students in learning mathematics. Laah-On, Intaros, and Sangaroon (2013) investigated problem-solving as an approach to help students learn mathematics. The researchers implemented a problem-solving process with an open-ended problem, self-learning, class discussion, and connecting ideas at the end of the learning process. Similarly, Jorgensen (2010) identified the need to improve students' curiosity and interest in math as a method to improve math skills. In this process, the student interacts with the instructor and other students to learn mathematics based on their experiences of working through the math problem. The researchers found that there were specific key universal activities in the problem-solving process with students. These key universal activities were counting, measuring, locating, designing building, playing, and describing. They concluded that these universal activities helped the students learn mathematics through the problem-solving process included in the classroom teaching methodology. Similarly, Schwartz (2012) found that having students interact in a math class with a game in a cooperative learning approach aided the students in learning mathematics.

SUMMARY CONCEPTS:

- Researchers have found connections between students' math achievement and their early math achievement in prior grades

- Another avenue for learning mathematics is with cooperative learning. Cooperative learning is defined as a method of studying in which students interact, discuss, and work together for math problem-solving

- Math instructors incorporate the use of videos to assist students in learning mathematics

- The key to math learning and teaching is for students to transfer mathematical understanding to new mathematical problems

- An effective model of teaching math is through the use of the flipped model of teaching, which includes the use of technology for lectures outside of the classroom and math assignments and activities as a focus within the math class

- Providing students a feedback process can enable students to use the feedback in the learning processes of mathematics

- Allowing students to construct knowledge through a hands-on approach of learning mathematics through problems that are anchored with real life problems assists students in their learning of mathematics

- When math instructors help students in challenging math steps, this helps students better understand the math concepts and math problems

Chapter Four

~~~

COLLEGE MATH EDUCATION

THE WORKFORCE IS experiencing high demand for college graduates with technical and mathematical skills. Furthermore, there is an increasing demand in the role of mathematics education reflected in real life mathematical problem-solving (Wake, 2014). The challenge for educators is to raise the technical and mathematical skills and knowledge for students (Mhakure & Mokoena, 2011). Moreover, college math education plays a role in this process because math educators are teaching mathematics and problem-solving procedures in the classroom (Zahner, Velazquez, Moschkovich, Vahey, & Lara-Meloy, 2012). In essence, a mathematics instructor in college can have a direct effect on a student's workforce skill development.

Several studies with effective teacher practices provide a recurring theme in math education. Wake (2014) reviewed a number of case studies where the key findings included allowing students to explore, use technology, and incorporate mathematical modeling in the solving of math problems. In contrast, another mixed-methods study with

842 teachers and 160 schools indicated that it was the lack of training and a lack of contextualized materials relating problems to real life that hindered the learning of math concepts (Mhakure & Mokoena, 2011). When math teachers slowed down in the lesson plan, and when they spent time in learning the math vocabulary, this helped students learn mathematics in the classroom (Zahner et al., 2012).

As a result, one of the focal points is to develop a student's conceptual knowledge in math. Students with a conceptual understanding had a strong foundation of the concepts and principles in math, and the students made connections to the content (Zahner et al., 2012). Additionally, students learn math concepts when they work together on problems. The researcher found a direct link to the discussion of mathematical problems in the classroom and the learning of mathematics.

Students are facing challenges in the completion of their college coursework. In a research study with 18 community colleges in California, VanOra (2012) found common challenges. The difficulty with college level coursework and curriculum inhibited college students from moving on to successful college courses in a community college or university (VanOra, 2012). Time was a major factor in prohibiting students from completing their college work because of time commitments outside of school. Pedagogy was another concern for students in completing college courses. VanOra indicated the inability of instructors to make real life connections to students because the material inhibited the students' comprehension of the coursework and material. In direct contrast, Varsavsky (2010) indicated in a study at Monash University in Australia, students with little math background from high school were just as successful as those students who obtained a high level of math in high school. *This might imply motivational factors could play a larger role in learning math than what was previously believed.*

Students see remedial mathematics as an irritating barrier in moving to college level math courses. Further, George (2010) investigated remedial math education in the U.S. from an ethical perspective. He saw remedial education as a barrier to higher education and college courses since many students could not pass the remedial math required. Taking this a step further, Varsavsky (2010) concluded that students with little math preparation in high school were as successful as those with higher levels of high school math preparation. This could calls into question the need for remedial math education in college. Students in remedial mathematics study algebra, geometry, and basic math, while attending college before moving on to college mathematics. Researchers argue that both instructor and administrators should use an ethical perspective when evaluating, conducting, and working with students in remedial college mathematics because remedial mathematics is seen as a barrier for many students (Bahr, 2012; George, 2010; Howell, 2011).

George, from a different perspective, concluded that if remedial college math were to be a part of college, then a standard exam should be used to enter and exit remedial math courses. In addition, Howell (2011) researched the California State University (CSU) remedial system in place and argued that the costs alone are enough to question the effectiveness of remedial math education. The CSU system spent $30 million in 2001 for remedial education, and the need for remediation could be a result of a student returning to school after an absence from high school. A student may take remedial classes because they did not learn the math skills required for college math while attending high school.

For many colleges and universities, developmental math education is a necessity for educating students. Bonham and Boylan (2011) did a descriptive review in their research of remedial education in the U.S. Three-quarters of U.S. colleges and universities offered a developmental math program (Bonham & Boylan, 2011). Bonham and Boylan indicated that 60% of colleges in the U.S. offered between two and four developmental math programs. Many college students entered developmental mathematics, and with assistance, only 10% of those students went on to graduate from college (Spradlin & Ackerman, 2010). Furthermore, Waycaster (2011) found that 14% of two-year colleges and 25% of four-year colleges evaluated and tracked developmental college students. Few of the students completed their developmental coursework to move on to college level courses (Hern, 2012; Sheldon & Durdella, 2010). Furthermore, Bonham and Boylan concluded that the high failure rates in remedial math education eventually caused students to withdraw from college.

There are contradictions in college developmental math education regarding the effectiveness of developmental math. For example, Bonham and Boylan (2011) saw developmental math as a barrier to students gaining access to college courses. Furthermore, Hern (2012) noted that math students who were required to take three or more developmental math courses in college had only a 10% success rate of completing a college level math course beyond their developmental college math sequence. Hence, many educators question the effectiveness of using developmental math education in college if there is such a high failure rate. Researchers believed the remedial math education system became a barrier to reaching college math, which is the opposite effect intended by colleges (Hern, 2012; Howard & Whitaker, 2011). From another perspective, rather than emphasize developmental curriculum, Albritton et al. investigated the use of an academic center with intensive tutoring at Seminole State College in Florida. Their results showed an increase of completion rates in developmental education by 15.5% and a return on investment of 272%. Additionally, results at an Australian University showed that students with little math preparation could be successful in college math, which contradicts the need for developmental math (Varsavsky, 2010). In short, although there are economic benefits for colleges offering developmental math, the remedial math courses are potential barriers for students seeking to move on to college math courses.

Instructors are using different techniques and methods to overcome these barriers faced by students in their completion of college math requirements. The research from VanOra (2012) showed the challenges students had persisting through multiple college math remedial courses. In two different studies where this writer researched the use of math in context, he incorporated the use of an online computer platform to teach mathematics for students who struggled. Utilizing programs such as Excel and online calculators, the writer was able to help students see the math they worked on in class. They also worked on mathematical problems related to their field of study. Results indicated that there was no difference in mathematical learning by simply using computers in the classroom, and there were significant gains in mathematical knowledge by using math problems related to the student's interests and field of study. In a qualitative phenomenological study on unsuccessful and successful experiences of learning math, Howard and Whitaker (2011) demonstrated that successful students had instructors who provided the relevance of the math they learned to their lives, and the instructors helped the students with their homework in class.

Math educators are identifying new ways of developing curriculum to aid students in accelerating their learning of math to address the barriers students are facing in completing college math. Many of the studies, which identified new approaches for teaching math, focused on the application of math or teaching of it in context.

One example is a qualitative study at a New Zealand high school that used video tapes as the data-gathering element. Harvey and Averill (2012) videotaped the lessons of an instructor with 22 students in the high school math class. The math lesson involved the triangle lengths in a Bailey bridge. The discussion of lengths of a triangle would have been very boring; however, the instructor used a real life object in the math curriculum, as in the Bailey bridge, to bring the math to a real life setting.

Results of the study showed the use of teaching in context assisted the high school students in learning math. In addition, new curriculum models such as accelerating the learning of math, integrating the learning of math in context, or reducing the number of remedial math courses helped to address the failure rates in college math and addressed the deficiency of math skills of students upon entering college (Asera, 2011; Valenzuela, 2012, 2014).

One of the goals of math education is the identification of effective math pedagogy and reform of math curriculum. Many educators and researchers see a need for math education reform. The research from Lundin (2012) indicated the need for math education to reflect the usefulness and real-world problems in math curriculum. Similarly, Merseth (2011), in researching the reform of a college curriculum, identified two shortened curriculum sequences with contextualized math adapted to real-life problem-solving. The first sequence, Statway curriculum, was developed for a shortened college statistics pathway, and the second sequence, Quantway curriculum, was reformed to shorten the college math quantitative reasoning requirements. The reforms in these curricula did not solely focus on the math problem. Additionally, the curricula emphasized productive persistence to support students through challenging math problems, and the faculty helped students with their homework in class. Researchers found the use of non-traditional pedagogy was an effective math-teaching method when compared to the traditional lecture and problem-solution teaching methods (Merseth, 2011; Vasquez Mireles, 2010). Vasquez Mireles (2010) researched a different reform effort that emphasized instructor training and development. The faculty and administrators at Texas State University in San Marcos reformed their developmental math sequence to include computer instruction in the curriculum along with mathematical modeling. The results of the research suggest there was an increase in students who passed the developmental math and college algebra because of the math curricula reforms.

However, ambiguity exists about what educational reform means and how to implement it in different schools. Some researchers do agree that there is a need for math educational reforms (Marz & Kelchtermans, 2013; Merseth, 2011; Vasquez Mireles, 2010). Educators and administrators could choose curriculum reforms based on value-based preferences or specific learning preferences such as through the incorporation of multiple intelligences as described in the Theory of Multiple Intelligences (Gardner, 2006, 2011). Further, a curriculum reform could be based on computer integration of math problems in and outside of the classroom (Vasquez Mireles, 2010).

From an entirely different perspective, the reformed curriculum could include an overlap of mathematical instruction with a content specialist to provide students with mathematical learning in their college content course (Valenzuela, 2012, 2014). Some researchers and educators saw the need to accelerate and shorten the sequencing of the college math curriculum for students to succeed (Perin, 2011; Vasquez Mireles, 2010; Valenzuela, 2012, 2014). Additionally, Marz and Kelchtermans (2013) concluded that math teachers' beliefs about math had an impact on the implementation of curriculum changes.

Math curriculum and instruction are two of the identified components to reform math education at the college level. Specific curriculum and instructional changes recommended to improve the completion rates of students for college math were accelerated learning programs, contextualized math, integrated curriculum, and support for developmental math learners (Merseth, 2011; Valenzuela, 2012, 2014).

Studies in accelerated and contextualized curriculum followed a similar instructional and curriculum approach as in the writer's own research. In this context, he developed a shorter math sequence to college math for students in a technical program at Lake Washington Institute of Technology in Kirkland, Washington. By allowing students to master modules in a course, students could accelerate through a given math course. The writer built the instructional model with context-based mathematical problems, which helped describe the technical concepts the students were learning in class. The math concepts covered in the class included mathematical concepts of electrical math, physics, computer math, and problem-solving. By the inclusion of context-based math problems in this curriculum, students were able to better grasp the math concepts because the problems were developed in applications of the technical fields the students were studying.

Merseth (2011), a Senior Advisor to the Carnegie Foundation for the Advancement of Teaching, described the 2009 development of the Statway and Quantway with the goal of doubling the number of students moving on to college math courses via the use of contextualized mathematics. However, Jorgensen (2010) noted that it was not only instruction in the classroom which aided students in learning, but student class participation and class attendance contributed to math success as instructors provided relevance to math in the real world as a theme to reform mathematics. The relevance to real world applications of math provided the motivation for students to continue learning math (Jorgensen, 2010).

To gain a complete picture of the non-completion of math requirements and the barriers to completing college math courses, a review of the college math placement systems may be a needed reform. The college math placement system was a typical math placement system in all colleges and universities. Most colleges utilized COMPASS as the placement testing system for English and math (Goeller, 2013).

Another college math placement system was called Accuplacer (College Board, 2015). The colleges and universities ensure that all students entering college are placed in the appropriate math and English course based on their level of knowledge through placement systems (Goeller, 2013). Goeller used a mixed-methods study to examine student perceptions of COMPASS, which is a college math placement system at a southwestern community college campus. Participants in the survey were from a developmental mathematics course at the community college selected from four sections of the course over four separate quarters. The researcher collected qualitative data through interviews, survey data, and the course objectives. The college also developed a pretest as a comparative tool to COMPASS for students in basic mathematics. Goeller (2013) found that 72% of the students felt they were in the correct class level after the placement test. Students also felt there was not an overlap of content from the COMPASS test to the developmental math course required. Furthermore, Venezia and Jaeger (2013) provided an insightful perspective when they indicated that the COMPASS and Accuplacer placement exams were poor predictors of success in college mathematics.

Developmental college math education is one of the methods for addressing the lack of college readiness. In some research studies, there is disagreement as to what college level readiness means in regard to student readiness for college coursework (Bailey, Jeong, & Cho, 2010; Goeller, 2013; Venezia & Jaeger, 2013). Developmental college math education aims to address the gap in mathematical background and knowledge with additional coursework in college before entering college-level math. Coursework in developmental math may include basic math and arithmetic up through pre-algebra. Bailey et al. (2010) provided a thorough review of the state of developmental college math. Bailey et al., in their regression analysis of the research databases, included data from over 57 colleges and 257,672 students. The researchers compared the data with the National Educational Longitudinal

Study of 1988. Their analysis and research showed that fewer than half of the developmental math students complete the developmental math sequence. Furthermore, their studies indicated that 30% of students referred to a developmental course did not take any remedial courses. Research results showed that, for community colleges, more than half of the students enrolled in at least one developmental course during their college career. They also indicated a cost of $1.9 to $2.3 billion dollars for developmental education at community colleges. One of the conclusions of the research included an emphasis on accelerated learning through the developmental math sequence.

Researchers are finding placement in the developmental math sequence is a barrier to college math completion. Colleges and universities placed students in college math remediation based on their math placement test scores (Bahr, 2012; George, 2010; Howell, 2011). Bahr (2012) provided further analysis and research on how the placement in a developmental math sequence affects college students. Bahr utilized a quantitative method of research, which included analysis of data and comparisons with the incorporation of regression analysis. Bahr analyzed and collected data from a database of the California Community College system. The investigator studied first-time college students over three semesters in the fall semesters between 2001 and 2003. The study consisted of 191,642 first time students enrolled in a remedial math course.

One of the findings included the observations that students beginning at lower entry points in the developmental math sequence had a higher probability of delaying their math sequence than those who entered at a higher-level of math. Furthermore, non-passing grades greatly affected their continuance in the math sequence. Moreover, students who faced additional steps in their sequence coursework had a higher chance of attrition than those with few developmental courses. To address these barriers, educators and researchers are recommending the development of new pedagogical and curriculum reforms to college math by using accelerated learning or contextualized math (Merseth, 2011; Perin, 2011; Valenzuela, 2012, 2014). These new methods of delivering math would assist students in reaching college level math (Perin, 2011; Valenzuela, 2012, 2014).

Howell (2011) noted that other issues besides the placement of students in a developmental math sequence were a barrier to moving to college math. Howell indicated that experienced teachers had lower remedial math courses taken by college students. Teacher retention was a key attribute to consider when students were not succeeding in developmental math. Moreover, students successful in high school with a 3.1 GPA found themselves in remedial education in the CSU system. These studies reflect some of the complexities of students not succeeding in their college remedial math or college math. *The curriculum may be of concern, yet there are other factors to consider such as the quality of math-teaching.*

The modern math classroom is no longer a face-to-face setting, and with an influx of new technology available for teaching, there is an increase in the use of online mathematics courses in college. Traditionally, most math courses are taught in a face-to-face setting in which students can interact with the instructor and ask questions in a real-time environment. Researchers, educators, and administrators are investigating the engagement levels of students in online math courses because of the prevalence of such modes of teaching in college. Petty and Farinde (2013) utilized content analysis to study the levels of engagement in an online class. There were 22 students in the study of a university methods class on mathematics. Six instructors were chosen from a middle school and high school. These taught algebra, calculus, and geometry. The instructors videotaped lessons, which were viewed by the research participants. The researchers investigated the extent that online math students were able to engage and understand math-teaching pedagogy in an online environment. Students were able to interact with the class through asynchronous discussion forums, synchronous communication online, and live classroom observations. The results from the researchers showed a higher percentage of engagement in strategy and clarification, and synchronous communication reflected a higher percentage of assessment engagement.

Math instructors have found differing results when comparing online versus face-to-face teaching of mathematics. Because there are contrasting methods for incorporating online learning in math courses, a research study conducted by Ashby et al. (2011) highlighted the teaching of math in a blended, online math, and face-to-face environment. The blended math course included both face-to-face and online components. Ashby et al. used a quantitative research method in the study with 167 students. The emphasis of this research focused on the success and differences in studying a developmental math course at a community college in three environments. The environments included online, blended learning, and face-to-face learning. The course chosen to study was an intermediate algebra course, which was one of three courses in a series in developmental math at a Mid-Atlantic Community College. Students in the blended learning class met in class one day a week and online the remainder of the week. *The results of the study revealed that each of the learning environments' success rates were not significantly different from each other.* However, when investigators considered completion, the face-to-face students did not perform as well in the blended and online environments. In short, the results from Ashby et al. mirrored other researchers' results in teaching math online (Mativo, Hill, & Godfrey, 2013; Schwartz, 2012; Wenner, Burn, & Baer, 2011).

In a contrasting view, Jones and Long (2013) noted a different result about online math versus face-to-face math education. The researchers used descriptive statistics to compare and analyze course grades in the online and face-to-face learning environments. *Seven out of the ten semesters showed no significant differences in the grades obtained from the online classes compared to the face-to-face math courses.*

Jones and Long concluded that the first three-semester courses had a higher percentage of passing grades for face-to-face math courses. The researchers conjectured the difference in scores due to the newness of online education at the college. The research by Jones and Long indicated challenges with comparing the two types of learning models for math. Furthermore, the research does not address the particular instructor differences in the class settings, which could affect the students passing the courses in the different models of math education.

Self-efficacy is an area of research within both education and math education. Self-efficacy is a social cognitive theory (Bandura, 1997), and it is becoming a part of the teaching theory in the classroom. Bandura's (1997) research on self-efficacy, as a subset of social cognitive theory, is based on cognitive, affective, biological, and behavioral factors in an individual. Bandura felt that self-efficacy was driven by accomplishments, vicarious experience, verbal persuasiveness, and the physiological. Moreover, from a math perspective, motivation has been shown to be a key attribute when learning mathematics.

Researchers are now in the midst of studying the role of self-efficacy factors in learning math (Adediwura, 2012; Erlich & Russ-Eft, 2011; Joet, Usher, & Bressoux, 2011). Levpuscek, Zupancic, and Socan (2013) completed one of the studies, which linked an improvement of math self-efficacy to achievement in math. Levpuscek et al. used a paired sample *t*-tests in the analysis of the different study factors. They researched math achievement through self-efficacy, personality traits, and parental support. The sample of the study included 386 students in the initial sample from the eighth grade, and the second part of the study included 372 students in the sample. The students were from 13 randomly selected public schools in the U.S. The researchers collected data over a two-year period with specific measurement tools to measure intelligence, personality types, student perceptions, and motivations. Results showed that students' intelligence and conscientious factors had a positive impact to self-efficacy in math, which had a positive effect on math achievement. There was a negative effect on self-efficacy when the parental pressure was considered and perceived by students. *The students' beliefs in their intelligence and the learning climate attributed by the instructor created a positive link to self-efficacy and their ability to do mathematics problems.*

Concerning self-efficacy, researchers are examining the role that gender plays in self-efficacy and the learning of math. Joet et al. (2011) provided a thorough review of this topic. The goal of their research study was to identify the relationship between self-efficacy and the learning of mathematics. Elementary schoolchildren were chosen because self-efficacy starts to evolve quickly in grade school children. There were 395 students from 19 different schools in France within this study. The researchers studied the effect of self-efficacy on academic learning for elementary school students. Moreover, self-efficacy was identified as the belief that students could accomplish a specific task or solve a problem, and students with higher self-efficacy have been shown to have higher persistence factors and higher achievement rates in school.

The researchers showed that girls had a lower self-efficacy in relation to math when compared to boys. They also revealed that self-efficacy through mastery learning improved achievement scores in mathematics learning. Some of the limitations of the study included not having a clear idea of the total impact to improving self-efficacy, the effect the teacher plays in the classroom, and the questionnaires not addressing the changes in self-efficacy. Yu-Liang (2012) studied fifth graders' mathematics self-efficacy and mathematical achievement and found that there was no significant difference between the math self-efficacy between boys and girls in the study. The researcher noted that a supportive parenting style had the best effect on math self-efficacy. Furthermore, there was a direct effect on the fifth graders' math self-efficacy and their math achievement.

The instructor has a direct impact on students' learning inside the math classroom whether in an online or face-to-face class. For instance, the instructor can motivate and engage students at different levels in a math course. Moreover, the instructor chooses the problems to cover in class, and he or she decides on the method of delivery. George (2010) concluded that the remedial math instructor in a college environment played a gatekeeper role for students because of the challenges students face in completing their remedial math requirements. Depending on how the instructor implements the problems and pedagogical methods in the remedial math course, the students were either motivated or discouraged in the classroom.

In the writer's own contextualized math curriculum, he mirrored an example of contextualization by developing context and real life math problems into an automotive technical program at Lake Washington Institute of Technology. He provided a supportive learning environment along with mathematical problems, which were of interest to students in their program of study. This pedagogical approach provided a positive learning environment for students, which was initiated by the math instructor. This is one of the reasons for the focus on researching contextualized math curriculum to improve a students' ability to learn math successfully (Bellamy & Mativo, 2010; Bottge & Cho, 2013; Harvey & Averill, 2012).

Several current pedagogical practices in math education are relevant to understanding how students learn. For instance, one of the learning processes involved in developing self-efficacy was the use of self-assessment to improve students' math skills (Adediwura, 2012). Adediwura studied the effect of self-assessment and self-efficacy with learning in mathematics. In the study, the researcher found motivation to be a key attribute when learning mathematics. Students' perception of themselves is just as important. As students' self-perception declined, their difficulty with learning math increased. Furthermore, in the study, self-assessment was defined as the involvement of students in the assessment and evaluation of their work. There were 60 total high school students randomly selected for this study. Adediwura noted that one of the key findings of self-assessments were that it did help improve student's self-efficacy about learning math. Students build self-efficacy when they feel they can persist through problem-solving (Bandura, 1997). Math instructors using self-assessment and self-reflection as a pedagogical tool in the math classroom aided students with their math problem-solving (Adediwura, 2012).

Memory has a role in the study of math and the learning of mathematics. Miller's (2011) research gives meaning to how educators and researchers can incorporate learning processes into long-term memory. Miller researched the areas of memory and its application to cognitive theory and learning. Miller's focus was developing an overview of the most recent cognitive theories of learning. The investigator viewed memory as sensory, short-term, and long-term. Each section of memory works together to either store or retrieve information. Recent research on phonological memory, as indicated by Miller, showed the use of this type of memory for vocabulary learning. Furthermore, Miller indicated that anchoring long-term memory to a context in different settings provided a better opportunity to access that information in the future. Short-term and working memory were not as critical to learning as were the engagement and involvement of students in curriculum lessons. Miller encouraged instructors to have students retrieve information frequently during lessons and in various methods and formats. By further analyzing the research done by Miller in the context of learning math, instructors focused on the memorization of mathematical formulas are emphasizing short-term memory for students. In contrast, Vanags, Pammer, and Brinker (2013) indicated that teachers who incorporated process-oriented learning through exploration, questioning, and an application of the material aided students in retaining what they learned in long-term memory. Instructors are encouraged to have students retrieve information frequently during lessons in various methods and formats.

Many educators are focusing on understanding the role of memory in math education. For instance, psychologists in the cognitive field had differing perspectives on the use of cognitive theories via pedagogical teaching methods because of the oversimplification of how memory worked (Miller, 2011). Math educators are researching new learning theories for students to retain and keep mathematical concepts learned in the student's long-term memory. One example was a math instructor's use of contextualized math curriculum where math was taught in the context of a real-life problem that students could relate to in their lives or future careers (Bottge & Cho, 2013; Valenzuela, 2012, 2014). Bottge et al. (2014) studied the error patterns with fractional computations between middle school students who received Enhanced Anchored Instruction (EAI) and those who did not. Key findings of the study reflected that the students with EAI reduced their computational errors and improved their math performance. The anchored instruction consisted of additional support to students through video instruction, face-to-face instruction, and emphasis on an applied problem such as building a hovercraft to pique a student's interest in mathematics. Furthermore, researchers found that it may be possible to increase numerical competencies through practicing math in board games, games with physical activity, or tracing number lines (Weigmann, 2013). In short, math instructors can assist students by emphasizing longer term memory usage through connecting math problems to various applications for students.

Math educators are examining how students learn math conceptually to understand the learning of math. In fact, for Khiat (2010), one of the central points of math education research was to comprehend how students' acquired conceptual knowledge of mathematics and how it changes from grade school up through college. The conceptual learning of math was where the students took the concepts of math and applied them to other problems or real life situations such as the study of an engineering application (Khiat, 2010). According to Khiat, math conceptualization theory was categorized into specific areas of learning: functional, procedural, disciplinary, and associational. The integration of each of these areas into the learning process was designed for students where the functional learning from students was based on how math could be applied (Khiat, 2010). Math educators associate an algorithmic form of learning with memorizing rules, definitions, and mathematical formulas. In addition, algorithmic learning was defined as a step-by-step procedure for solving a mathematics problem.

Summary Concepts

- There is an increasing demand and role of mathematics education reflected in real life mathematical problem-solving

- Allowing students to explore, use technology, and incorporate mathematical modeling in the solving of math problems helps students increase their math knowledge

- Instructors can incorporate the use of an online computer platform to teach and learn mathematics for students who struggle in math. This is much more prevalent with online learning systems incorporated in online and face-to-face math classes

- Successful students have instructors who illustrate the relevance of the math they learned, and connect the math to their lives, even helping the students with their homework in class

- Students must be helped to develop productive persistence by math faculty through challenging math problems in which they provide aid

- With technology, studies show that each of the math learning environments online and face-to-face success rates are not significantly different from each other

- Math instructors need to bring the math concepts to life beyond the use of mathematical symbolism through real life applied problems

Chapter Five

Linking Mathematics and Learning

One of the obstacles in learning math is that students lack the application of mathematics to their own lives (Perin, 2011). Many math educators and researchers have learned to link contextualization to other math education processes such as integrated learning, integrated curriculum, contextualized teaching, and work-based learning (Bottge & Cho, 2013; Perin, 2011; Valenzuela, 2012, 2014). The emphasis here will be to describe how contextualized math curriculum aids students in linking mathematical concepts, organizing math ideas, and solving mathematical problems.

Linking Mathematical Ideas

Researchers and educators are identifying how contextualized math curriculum assists students with the linking of math concepts. One of the prominent researchers in the field of researching contextualized mathematics is Brian Bottge. Bottge furthered the research on linking mathematical concepts through contextualized math curricula by working with researchers who have backgrounds in biostatistics, mathematics, and multimedia (Bottge et al., 2010; Bottge & Cho, 2013). The researchers utilized a randomized pretest-posttest research method to compare the pedagogical teaching strategies in math education. The instructors enhanced the classes with the use of a video describing the fractional measurements with a ruler. The researchers found students were able to construct and link math concepts and improve their math skills through math instruction when lessons were engaging.

Students can link mathematical concepts via problems of interest and modeling of real life problems. Two research studies provide additional evidence as to the linking of math ideas through contextualized learning. O'Shea and Leavy (2013) used a qualitative case study to identify how instructors developed a constructivist learning approach for math problem-solving. The study was primary school teachers with students 10 to 12 years old. There were five teachers in the study, and the researchers gathered data from interviews, focus group meetings, and observations. The researchers described constructivism through an active learning process of inquiry, problem-solving, and working with others on problems. Through their research, they identified effective learning when students engaged the problem-solving process, and instructors helped students in problem-solving by choosing problems of interest and those that pertained to students. In comparison, Doerr et al. (2014) investigated the use of a modeling-based summer bridge mathematics course and its effect on student achievement in their first-semester college mathematics course by using a quasi-experimental method. The researchers compared the first college mathematics course after the summer bridge class. The investigators selected participants who either attended a traditional first quarter college math course versus those who also attended the summer bridge program. The cohort comparisons were from 2007 to 2009 and 2010 to 2012. The emphasis of the course was to prepare students for pre-calculus and calculus in college. The modeling activities included rate of change problems and electrical resistor-capacitor modeling problems. The students were encouraged to work together, test their data, and come up with mathematical models to interpret the problems. *The researchers found higher-grade gains from the students who took the modeling-based course versus those who took the traditional math course*, and the modeling activities helped students link and develop mathematical concepts.

The incorporation of real world context is a component to aiding students in linking mathematical concepts in the classroom. Tosmur-Bayazit and Ubuz (2013) studied the use of engineering in mathematics education within a college environment. The investigators based the learning of mathematics on the material that was covered in the math class. The research emphasized the use and application of engineering math to learn both mathematics and engineering at the college level. Five engineers in the study developed electrical components for the Turkish army. The researchers collected data through interviews, and they analyzed the data through pattern matching. Three themes emerged from the research: the teaching of mathematics, analytical thinking, and mathematics in general. One of the findings is that mathematics courses should include engineering contexts to help students learn and develop math concepts. Tosmur-Bayazit and Ubuz recommended that math could help engineers interpret the physical world with mathematical models. Taking this a step further, the context of the applied engineering could aid students in learning math. They concluded that instructors of mathematics should include real world applications to the math problem sets students worked on in the classroom.

Instructors using authentic learning provide real world problems to students, and as we have seen repeatedly, this is an effective teaching method. For instance, Strimel (2014) described how to enhance authentic situations in education through real life situations. With education emphasizing Science, Technology, Engineering, and Math (STEM), there is a renewed focus on using authentic teaching as an avenue for students to delve into STEM topics. Authentic learning is similar to the use of contextualized math curriculum, with students working on problems based on real situations in society (Bottge et al., 2010; Perin, 2011, Valenzuela, 2012). One of the methods described by Strimel included the use of self-directed learning in which students identified the math problems to work on and solve. Strimel noted that students would be interested in solving problems they researched or developed themselves. The goal of authentic learning was to ensure the lesson that was developed added meaning for the students.

Furthermore, authentic learning has connections to the research and work completed by Bruner (1977). Bruner noted that learning should be a foundational basis from which to build on and transfer knowledge. By adding detail to a topic through simplification and added meaning, students will be in a better position to remember those topics. Similarly, Vygotsky (1978) saw learning and development as two processes: learning was first, development second. Learning was not just a thinking process, but learning was a way of independently developing the processes for further thinking. In contrast, Strimel thought authentic learning should include the students' lives and perspectives in solving the problem, and this frame of reference has connections to constructivism (Dewey, 2011; Piaget, 2001; Vygotsky, 1978). Strimel provided an example in the use of an earthquake to develop an authentic, non-structured problem for the students to solve. The students had three questions in their research and problem-solving, culminating in the applications to engineering and science. The students were also required to collaborate and work together on the problem. The researcher recommended the use of multiple subjects and topics within mathematics to solve the authentic problems. Strimel ended the process with time for the students to reflect on their solutions and present their findings.

Authentic assessment is another form of using math contextually to build upon the math skills of students in a classroom setting. Tammaro and Solco (2013) investigated the use of authentic assessment as a measure of gauging students on their application of new insights rather than just testing on what students already knew. Educators defined assessment as a method to gauge what a student knew at a point in time where the information gathered in the assessment allowed the instructor to change the lessons and curriculum if needed or evaluate the effectiveness of the curriculum (Tammaro & Solco, 2013). The investigators viewed authentic assessment as a measure of how students applied what they knew at a given point in time (Tammaro & Solco, 2013). In contrast, contextualized learning was thought of as a method of teaching math where students learned skills that would help them in the job market (Perin, 2011). These skills are developed through critical thinking and working on real world problems (Tosmur-Bayazit & Ubuz, 2013). In this respect, *the knowledge gained by students was equated with the skill development required in the job market.* This was particularly important for adult learners looking for work. In authentic assessment, the problems are contextualized, and students experiment to come up with real-world solutions. Students are using the authentic learning to construct and link concepts learned in class, which incorporates the use of constructivism (Dewey, 2011; Piaget, 2001; Vygotsky, 1978).

Authentic learning can also be an effective form of teaching in a math class. Busadee and Laosinchai (2013) studied the use of authentic learning via probability math problems. They investigated the use of authentic problems in probability, combination and permutation problems, at the high school level. Combinations and permutations are part of statistics, which provide for the counting of different groups of data whether the order is or is not a factor in the collection of data. The researcher's primary emphasis was gathering data on whether the authentic learning units could improve the mathematical understanding of the probability problems. In contrast, Prado and Gravoso (2011) utilized anchored instruction with real life problems in statistics anchored by a video to reinforce the statistical topics to the high school students, and they found anchored instruction to be effective in learning statistics. On the other hand, Busadee and Laosinchai used data obtained between 2008 and 2010 in pretest and posttest scores with three groupings of probability problems such as sports, word problems, and probability game problems. The researchers concluded that the students needed real life examples for problem-solving to assist the students in linking math concepts and understanding the math problems.

Authentic learning has similarities to an integrated curriculum. Lynott, Kracl, Knoell, and Harshbarger (2013) researched the use of interdisciplinary studies for an authentic learning experience. Educators defined interdisciplinary studies as two instructors working together to integrate topics for the benefit of the students. The investigators emphasized the authentic learning approach through an integrated curriculum method with interdisciplinary topics. In contrast to other studies on integrated curriculum, Ngu and Yeung (2013) studied the use of integrated mathematics in a chemistry class to help students understand the foundational math concepts in chemistry. The embedding of mathematics in chemistry aided the students with the foundational math concepts required in the chemistry course. On the other hand, Lynott et al. focused on physical education rather than mathematics. Lynott et al. developed an integrated curriculum using mathematics, science, and physical education for students to measure, demonstrate, and assess a variety of skills and concepts in these subjects. The students completed the mathematical assessment through graphs and analysis of the data gathered. Lynott et al. noted that the collaboration between instructors in the development of authentic interdisciplinary lessons was an essential element in the integration of curriculum, and they recommended developing one lesson per semester because of the time commitment involved. In contrast, Araujo et al. (2013), within their integrated curriculum research, *concluded that the real world concepts brought meaning to the students while they worked on the mathematical problems.*

ORGANIZING MATH IDEAS

The use of authentic math instruction, a form of contextualized instruction, benefits the students in the use of authentic lessons. Dennis and O'Hair (2010) completed a qualitative comparison case study through document reviews, interviews, and observations at three high schools on authentic lessons, and the real-world math problems added value to the student beyond an actual math test. One of the unique qualities of this study is that it highlighted the use of contextualization via calculations of iron content on breakfast cereals as a math lesson to assist students. There were three high schools and five math instructors in the authentic research study. The goal of the study was to see how authentic instruction aided the students in successful completion of math, and they studied the integration of authentic instruction through the math curriculum.

One of the findings in the qualitative study was the lack of time available for the development of authentic lessons. Attendance was also a challenge when teaching project-oriented lessons to students who missed critical material in class. The time required for developing authentic lessons and the amount of material needed to cover an authentic lesson was one of the disadvantages. Moreover, the use of the application problem of building a hovercraft aided the students in improving their math computations.

Math instructors improve the organization of mathematical ideas for students through the integration of non-traditional curriculum in college. In my work and research on contextualized mathematics, this writer focused on developing a math program that incorporated teaching math in context to technical areas in the automotive repair program at Lake Washington Institute of Technology. The math concepts highlighted within the course allowed students to comprehend math ideas about their technical program. Math topics included binary computations, series and parallel circuit calculations, hydraulic lift problems, and logic gate processing. The author concluded that students in the contextualized math courses had a 70% success rate and higher retention rates when compared to those who were not in the contextualized math courses.

Furthermore, in a research study comparing the use of context-based math curriculum for English as a second language (ESL) learners in college, the group in the contextualized class performed slightly better than those who were not in the contextualized learning where the problems were about finance, shopping, and balancing a checkbook in context (Valenzuela, 2014). Both of these math curriculum models are examples of teaching with contextualized math in the Integrated Basic Education Skills (I-BEST) model that includes two instructors in the classroom. As an example, one instructor was a content specialist teaching automotive topics, and an adjoining instructor, this writer, taught an academic topic such as mathematics. Furthermore, both the automotive and ESL programs incorporated online learning platforms so students could organize the math concepts online through discussion forums, online assignments, and video lessons.

Developing contextualized mathematics curriculum to help students organize mathematical ideas starts with the development of problems through the solving of which students could conceptualize applications. Wilson (2011), in his work on developing contextualized math problems, understood the key principles of having adults comprehend mathematics. Wilson, at the time of his research, was the Math Department Chair at an Arizona community college. In Wilson's research, he described his method of engaging 80 employees about the mathematics behind a companies' bonus plan workings. The audience learned specific math applications and concepts of linear modeling, exponential functions, and piecewise functional modeling. The example provided was an excellent example of contextualized mathematics in the real world. Wilson understood that by selecting a topic which the employees were interested in, and presenting it in a logical manner, it helped them understand and organize the mathematical computations behind the presentation.

Pedagogical practices are some of the tools used to aid students in organizing mathematical ideas with a contextualized math curriculum. Bellamy and Mativo (2010) described the importance of meaning and context in mathematics that students can relate to in real life. The emphasis of their research was providing pedagogical ideas for teaching students at the middle school level. Bellamy and Mativo's research detailed specific teacher practices and ideas for helping students understand mathematics.

One of the real-life applications of math was the use of the Pythagorean Theorem within the construction trade to find measurements of a given side of a right triangle by knowing the length of the other two sides. The researchers recommended inviting professionals into the classroom to add to the lesson plan subject as subject matter experts. Furthermore, by using a ruler as a measurement tool, and specific two-by-four inch pieces of lumber, students could investigate measurements and construction within the classroom. Bellamy and Mativo determined that the classroom could be a laboratory setting for exploration and learning for the students when studying mathematics and the sciences. Similarly, Fatade et al. (2014) indicated which students were engaged in math compared to those who were not. They did this by providing a problem-based contextual math problem versus a traditional numeral-based problem in the math classroom. However, conversely Battey

(2013) noted that other teacher attributes hindered math learning by ignoring the students' thought processes and using sarcasm in a negative manner in the classroom.

An instructor's use of modeling-based mathematics curriculum is another form of incorporating contextualized mathematics curriculum in the classroom. Budinski and Takaci (2013) described modeling-based mathematics as the use of equations or an algorithm to describe a specific situation in the real world. This is similar to providing contextualization to a real world problem with mathematics (Perin, 2011). Part of the teaching process for modeling-based math was to provide students with guidance on a topic and guidance on the data required for the problem. In this respect, the data and modeling are added context similar to the context that the contextualized Enhanced Anchored Instruction in the research by Bottge et al. (2010). One of the challenges in developing math-based modeling of problems was the time required to set them up. This was also a theme in the findings through the research identified by Mhakure and Mokoena (2011). On the other hand, Budinski and Takaci (2013) indicated that the use of computer spreadsheets aided students in developing graphs and graphical representations of data. One of the resources recommended was www.GeoGebra.org (Budinski & Takaci, 2013). This site provided teachers and students the ability to see math work dynamically in graphical- and modeling-based applications of mathematics.

The methods described by the researchers were integrated into the teaching practices of Serbian high school instructors teaching mathematics. One of the examples provided was the math modeling of an earthquake in Japan. The use of modeling of logarithmic data to compare the magnitude of two different earthquakes was the central problem. Researchers modeled data using the formula for magnitude via the application of logarithms, and they used Internet resources such as GeoGebra (The International GeoGebra Institute, 2015) as a resource to represent graphically the data modeled. In this process, students gave their reflective thoughts on the modeling and computation process. Budinski and Takaci's research revealed that the use of math modeling of a real world scenario aided students in understanding math and improved their problem-solving skills in math.

From a different perspective, Toews (2012) described the use of mathematical modeling in his undergraduate math course at the University of Puget Sound with 20 students. In addition to mathematical modeling, other aspects of the course included collaboration, use of the computer for modeling, and conducting research. The researcher used a mathematics modeling text in biology to bring context to the math topic. As part of the course, students were expected to do low-level computer coding in a computer language. Students were required to complete a 10 to 20-page final project report as part of the mathematical modeling class. The mathematical modeling took the form of using the computer to decipher a problem, analyze it, and then come to specific conclusions. Within the course, students had to complete presentations and write extensively about their homework and project. The investigator defined the modeling aspect as a mathematical method of describing a specific real-world scenario. The researcher found an important part of modeling was the data collected. An example provided was the use of sensor devices, which provided data involving an aerospace engineer for satellite configurations. Toews found that mathematical modeling provided students with real world work experience to be able to collaborate with others in a variety of disciplines.

SOLVING MATHEMATICAL PROBLEMS

Math instructors help students in solving math problems and learning math via a contextualized math curriculum. One of the key studies, Deed et al. (2012), highlighted the use of engagement as a method to aid students in solving math problems and learning math in a contextualized math curriculum. Deed et al. used a case study methodology that included seven students in the sample within a sophomore-year mathematics course. The researchers emphasized the experiences of the students while learning mathematics, and they collected data through observations of the classroom and interviews of students. The researchers analyzed the data through a coding process to identify relevant themes. One of the key findings was that students enjoyed learning math and solving problems that would be used and applied outside of the classroom.

Students used drawings to explain their learning experiences in the classroom. Furthermore, students needed a safe environment to be able to ask questions and experiment with their ideas. Engaging students with different approaches to math problem-solving helped students in the learning process. Similarly, Nijlen and Janssen (2015) concluded that students who were not confident in math were still motivated to work and solve math problems on a contextualized assessment, and there was a tendency of lower effort in non-contextualized assessment questions versus contextualized math questions. In essence, students in both cases were motivated to solve math problems because of the contextualized math material.

The curriculum and lessons within contextualized math curricula provide a basis for problem-solving. There were two notable research studies done by Weiss, Herbst, and Chen (2009) and O'Brien, Wallach, and Mash-Duncan (2011). The findings and perspectives from these studies provide further understanding in the problem-solving arena where students used contextualized curriculum. Weiss et al. researched the use of authentic mathematics to teach geometry where authentic math was another name for contextualized math. The researchers used a qualitative method in their research. They included videotapes of 26 teachers from 19 high schools, and they gathered data from five focus groups.

One of the findings of the qualitative study was that the teachers felt mathematical proofs assisted students in thinking critically about a "problem to solve the problem." The assumptions students took during the exercise in proving the problem mathematically were of equal importance. In the video and discussion afterward, the instructors viewed the different authentic mathematics applications from the students. The researchers felt the use of the two-column proof was effective in teaching students math. The students' enthusiasm for the problems helped them towards the solution of it. This reinforces the need to have a curriculum, which interests students and captivates their curiosity to lead them to math problem-solving.

SUMMARY CONCEPTS:

- One of the obstacles for students learning of math is that students lack the application of mathematics to their own lives

- When students are allowed to make mathematical concepts out of real life concepts such as fraction strips of paper, they can see what math looks like in real life

- Effective math learning with students includes engaging them in the problem-solving process, and instructors allowing students to choose application problems or a topic for a math project

- The teaching process for modeling-based math is giving students guidance on a math topic and guidance on the data required for the problem

- Engaging students with different approaches to math problem-solving helped students in the learning process

Chapter Six

Contextualized Mathematics Teaching

Contextualized mathematics teaching can be accomplished through problem-based learning in the classroom. Notable results with studies in problem-based learning (PBL) show promise in the areas of mathematical learning for students. Two notable studies reflect how PBL is applied and used in the classroom (Goodman, 2010; Trinter et al., 2015). Instructors teaching with PBL use inquiry, motivation, research, and active collaboration to allow students to work through real world-contextualized math problems to find a solution.

The conclusion was that PBL curriculum (i.e., contextualized math curriculum) emphasized problem-solving, communication, reasoning, and the connection of math concepts.

Chang, Huang, and Liu (2012) studied the use of contextualized math learning with computer-assistance. Moreover, Chang designed a computer-assisted system called MathCal to incorporate problem-solving steps in a math-contextualized curriculum for students to go through to solve a mathematical problem. A visual display of the mathematical problem along with additional questions asked by students allowed students to work through the problems with the computer system. The mathematical computer-assisted model included guidance on problem-solving, a student history record, teaching material, and interactive feedback for students. There were six third-grade classes from a county in Taiwan. Students were selected because they required remedial assistance in mathematics. Chang found that the students in the experimental group, students using the contextualized computer-assisted system, did better than students without the computer-assisted math problem-solving system. Chang concluded that the computer-assisted system, and math contextualization incorporated into the computer system, helped develop the problem-solving skills of the remedial math students.

Math educators incorporate mathematical modeling as a method of assisting students in their problem-solving. Mathematical modeling was described as the use of mathematics to bring context to a real life situation. In this sense, modeling is the use of contextualized math material for mathematics curriculum (Perin, 2011). Researchers indicated differences between traditional problem-solving and mathematical modeling. In traditional problem-solving, students utilized formulas or algorithms to solve math problems. In modeling, students worked together to review data and come up with a mathematical process to describe the authentic real life situation.

Other researchers found that collaboration and learning by working together was an effective method for learning math. Math modeling was considered an effective method for students to improve their math problem-solving skills. The two approaches for the use of math modeling assisted students in learning mathematics, and it aided them in the discovery learning-process within the math lesson. Erbas et al. identified two examples, which could be used as a form of this type of discovery learning. For instance, a math instructor utilization of logarithmic exponential model-equation provides students a description and understanding of how long it will take radioactive material to decay. In the second instance, students can inductively work on developing their equation to describe a pandemic disease, which spreads exponentially. One of the challenges posed by the researchers was the lack of resources for instructors to develop curriculum using mathematical modeling. On the other hand, Castillo-Garsow (2014) saw problem-solving as a process of having a task, problem, and solution. Castillo-Garsow argued that other researchers should emphasize research in the area of modeling and research modeling because his research showed that students began to develop their ability to problem-solve, abstract, and develop models based on the modeling assignments.

The integration of technology with modeling-based math curriculum is at the forefront of problem-solving. Lavidas, Komis, and Gialamas (2013) studied the use of spreadsheets as a method for solving problems in context. They used spreadsheets to solve algebraic equations in story problems. The story problems were another form of contextualized curriculum used by educators and researchers (Bottge et al., 2010; Perin, 2011; Valenzuela, 2012, 2014). Researchers used the computer spreadsheets as a method for problem-based learning in mathematical problem-solving. College students had to enter formulas into the spreadsheets to confirm their answers. Perin (2011) found that contextualized math problems were similar to the mathematical story problems and derived from everyday situations. Lavidas et al. found the cognitive processes in solving the math problem involved the use of prior knowledge to extract data from the contextual problem, and then apply a general equation to solve the problem. In a similar fashion, the spreadsheet is modeling the problem-solving process.

Similarly, other researchers in contextualized math studies used pretests and posttests in their research (Bottge et al., 2010; Bottge & Cho, 2013). Math researchers developed the calculation tools in MS-Excel, which included the average, sum, and functions within the MS-Excel application for calculating mathematical problems. In the experiment, students were given two problems. One of the problems was calculating tax on products purchased using an algebraic equation. The second problem was the development of an algebraic equation using the voting results for student body elections. There were 124 students in the study, and they were at the college level ranging from 19 to 20 years of age. Research noted that students verified their solutions frequently using spreadsheets in comparison to students who solved math problems manually. They further found that the students had an effective transition from arithmetic to algebraic problem-solving with the use of spreadsheets. Similar research showed that the contextualization and anchored instruction helped the students improve their computational math skills.

Anchored instruction, another form of contextualized math learning, is a method of learning problem-solving in the math class. For example, Gunbas (2015) researched the use of Computer Assisted Instruction (CAI) with mathematical story problems to assist students in learning and solving math problems. The math, in this case, was anchored instruction with a computer-based story problem. The word problem provided context to the math problem, which aided the students in solving mathematical problems. The story problems in the study reflected everyday scenarios with added math contextualization to the learning process in solving the math problem. The researcher focused on whether the word problem helped the students in problem-solving, and if the word problem in a CAI program had different results than in a traditional paper-based form math class. In Gunbas' study, students either solved the word problem on the computer or individually on paper. Gunbas noted that the posttest results were significantly higher in the student group using the CAI word problem context versus those solving the math problem on paper. Moreover, the anchored instruction of using the word problem in combination with the computer helped the student's problem-solving processes. This was similar to the results from other researchers (Bottge et al., 2010; Bottge et al., 2014).

Math instructors utilize anchored instruction in math courses such as statistics. Prado and Gravoso (2011) studied the use of anchored instruction in a statistics class. Anchored instruction is a teaching method placing emphasis on real-life problem-solving to bring meaning to the statistical problems. This is similar to the contextual math curriculum in other studies (Perin, 2011; Valenzuela, 2012, 2014). In Prado and Gravoso's research, the instruction was anchored by a lesson plan designed to have students answer statistical distribution questions by watching a video and reading a story to explain the concepts. The participants in the study were second-year high school students in Leyte, Philippines taking an introductory statistics course, and there were 64 students in the research study. As a comparison/contrast in the study, another class did not use the anchored instruction.

Prado and Gravoso found that students in the anchored instructional learning model had higher achievement scores on the post-tests than those students who did not receive the anchored instruction. This was similar to the findings by Bottge and Cho in their research on anchored instruction. Prado and Gravoso concluded that the realistic scenarios the instructors exposed the students to in the anchored instructional classes helped them in the learning and problem-solving process. In addition, they felt that the anchored problems provided students a collaborative environment to learn and work together. The collaborative nature of anchored instruction was similar to the problem-based learning collaborative environment found by O'Brien et al. (2011) in their investigation of problem-based learning.

Shiu (2013) studied a new method of anchoring real life data from students' lives in the teaching of a statistics course. Students answered a series of questions about their lives to gather data, which utilized a form of constructivism in the course. Shiu utilized a pilot study to integrate this new teaching method for a statistics class. The researcher gathered the data from all students and developed a spreadsheet. From the spreadsheet, the researcher was able to develop statistical problems, which included the real-life data from each of the students. There were 110 students in the study in the undergraduate statistics course. Students attended their class in a lab setting with access to computers and statistical spreadsheets. Shiu revealed through the new pilot course the fact that students' enthusiasm and interest in learning statistics increased because of the connection to the data.

Understanding the comparison of traditional classrooms with those having anchored instruction provides researchers and educators knowledge of how anchored instruction works in math courses. For instance, Elcin and Sezer (2014) studied the comparison of anchored instruction in a math class setting. The researchers saw anchored instruction similar to problem-based learning. Both of those forms of learning are a form of contextualized math curriculum (Perin, 2011). The instructors anchored the math problems to real-life problems illustrated in a video. Elcin and Sezer's study consisted of two courses in the sixth grade with instructors teaching mathematics to students in Istanbul, Turkey. The researchers revealed there was a significant difference in the posttest scores where the anchored instruction class did better than traditional math class instruction. Furthermore, the anchored instruction information was retained longer than in the traditional form of teaching. In a similar fashion, Miller (2011) indicated that this type of anchored methodology improved the recall of information on his research regarding memory.

The use of computer software is another method of incorporating anchored instruction in math classes. Zydney, Bathke, and Hasselbring (2014) investigated the use of anchored instruction with real world problems in the context of a software program, which included videos to help students solve mathematical problems. The researchers wanted to identify the appropriate level of assistance students needed in the anchored instruction environment. One of the typical real-world problems related to determining if an iPod had enough space to hold a song. Students then had to do the mathematical calculations for the memory space required.

Another example was the conversion of units to the metric system of measure in baking a batch of cookies. The study included two classes of fifth-grade students. The instructors completed the comparison through two versions of the computer software where one version covered the basic concepts up front and the second version emphasized structured problem-solving. The computer software included helpful videos and animation for students to follow. The use of the computer was similar to the use of the Computer Assisted Instruction study by Gunbas (2015) where the anchored instructional method relied on the computer system. The researcher found that the class with the structured problem-solving did better in math problem-solving than the basic concept class. The structured problem-solving group used a comparison feature, which compared similar types of math problems and the steps involved in solving them. However, when the math problems became complex to the extent that they required multiple operations to solve, there was no significant difference between both groups, which differed from results found by other researchers (Bottge & Cho, 2013; Gunbas, 2015).

Embedded math is another form of math contextualization. For example, Burn, Baer, and Wenner (2013) studied the use of embedded modules of mathematics in geoscience courses. The embedded math modules were needed because students often lacked the quantitative skills for the mathematical rigor in geoscience courses. In 2010, the author developed an embedded mathematics course in an auto repair program at Lake Washington Institute of Technology. This writer developed math modules to help students in understanding how mathematics related to the auto repair technician courses. He integrated in the mathematics curriculum topics related to physics, electronics and computer mathematics in the auto repair program.

Vasquez Mireles (2010) noted that the use of modeling of mathematical data aided students in the linking of concepts and the solving of math problems. Moreover, Ngu and Yeung investigated mathematical examples demonstrating the effective steps needed for students to construct the math solutions to similar problems. There were 22 students from a Malaysian high school in the study. Students were in two classes over a two-day period where they studied the worked examples method or the text editing problem-solving technique. Both were in a chemistry class. The researchers used analysis of variance and descriptive statistics to analyze the results. Ngu and Yeung concluded that the group of students in the worked examples class did better at solving the mathematical problems than students in the chemistry class using text editing. Some students were also able to solve mathematical problems using a two-step strategy.

In addition, Kim and Cho (2015) examined the use of integrated curriculum and design in science and mathematics courses. The researchers found that the integration of both math and science aided students in understanding real world concepts through integrated instruction. They also revealed that constructivism could be an integrated element in course design by building in lesson plans, which invite exploration and active participation by the students. Kim and Cho used a case study research methodology. There were 37 students in the study at the grade school level within the city of Seoul. In a similar fashion, Gucler (2014) utilized a case study to research a college-level calculus course with 23 students where the case study format allowed Gucler to provide insight and understanding of the math-learning processes to his students.

The students in Kim and Cho's research were provided an integrated lesson plan on the exploration of symmetry by using a mirror. The mathematical concepts included the use of symmetry and transformation and the science behind the mirror was the use of reflected light. Kim and Cho found that the integrated curriculum of both math and science helped students gain interest in the lesson plan and motivated the students to learn about those topics. The integrated lesson provided students an opportunity for cooperative learning, and students learned through experimentation.

Research is underway to identify the Common Core State Standards emphasis in an integrated curriculum. Researchers Schwols and Miller (2012) found key issues related to the need for remediation of math for students in science classes. An example of the use of integrated math was the kinetic energy component in a science class, which could include a mathematical module on the math behind kinetic energy. The use of integrated math in a science course is a means to improve a student's interest in math and improve course mastery such as in the work by Erlich and Russ-Eft (2011) on social cognitive learning. Schwols and Miller's recommendation was that *science faculty collaborate with mathematics faculty to integrate the necessary math content to support students in their science programs.* For instance, one of the Common Core Math Standards for eight grade students stated the need to understand exponents. Schwols and Miller recommended using an integrated module on teaching students exponential growth from a math perspective to support their knowledge in science. Although there was not a rigorous study done on the use of integrated math, the benefits and similarities are found in other research on embedded and integrated math curricula (Perin, 2011; Valenzuela, 2012, 2014).

Educators and researchers improved their understanding of integrated curriculum via the different elements and implementations of an integrated curriculum. Araujo et al. (2013) investigated instructors' conceptions of integrated math curriculum within the context of a new curriculum implementation for instructors. The researchers viewed conceptions as the instructor's understanding of the goals and implementation of the integrated curriculum. Researchers and instructors viewed integrated curriculum as having connections with other branches of mathematics and other content areas such as science (Araujo et al., 2013). They described the strands as the different branches of mathematics such as algebra and geometry. Additionally, integration by topic allowed students to work on math problems and explore different elements of math. Instructors could also implement the interdisciplinary, integrated math concepts in content areas such as science, vocations, or the arts.

Summary Concepts:

- Contextualized mathematics teaching can be accomplished through problem-based learning in the classroom. Problem-based learning can be provided in an online and face-to-face setting

- The use of mathematical computer-assisted modeling and online help features provide students with immediate assistance to help build their confidence and learning of math

- The new technology offered to students in our online and face-to-face classes need to also provide students with immediate feedback so they can make real time connects to math concepts they are learning

- Using computer spreadsheets as a method for problem-based

learning in mathematical problem-solving helps students learn math as they are entering formulas in spreadsheets like Excel

- Math instructors can integrate math curriculum with connections in other branches of science or other technical areas to enhance math learning of students. The integrated math curriculum can draw math concept connections to other scientific fields at the college and university levels

Chapter Seven

A New Mathematics Curriculum

THERE IS A need to incorporate new curriculum in the college math classrooms because of the challenges college students have in passing their college math classes (Albritton et al., 2010; Bonham & Boylan, 2011; Hern, 2012). Through the use of college contextualized math curriculum, students are able to link mathematical ideas, organize math concepts, and successfully solve mathematical problems in the research study. By incorporating college contextualized math curriculum in the classroom, math faculty and administrators will provide an engaging class environment for students (Toews, 2012). Moreover, through this engagement of mathematics via real world problems, students will further explore and decipher the mathematical content in the contextualized math problems (Prado & Gravoso, 2011). Further, the use of college contextualized math curriculum can be applied in the developmental college mathematics level (i.e., pre-college mathematics) and in the college mathematics level (Perin, 2011; Young et al., 2012).

When math instructors use a contextualized math curriculum it assists students in the learning of mathematics. First, through the use of a contextualized math curriculum, students are more engaged in class with real world problems. The real world math problems also assist students in linking mathematical concepts in the classroom. When students develop their conclusions based on the contextualized math, this aids them in linking concepts, and students are then able to see patterns in the real life math problems, which assists them in further linking mathematical ideas. Furthermore, when students have real life math problems to work on, it assists them in organizing their mathematical ideas. Also, the linking of math concepts to prior math concepts learned in contextualized math problems, helps students organize the math concepts. By breaking down the real life problem into mini-problems, the students are better able to organize the mathematical problems and solution steps. Furthermore, when math faculty facilitate the learning process with contextualized math, it helps math students solve the problems successfully. Facilitation in this sense relates to math instructors connecting the math lessons to the resources and support for students. In addition, having a safe place for students to work on contextualized mathematics is important in the solving of mathematical problems. A secure environment for math students is an environment where they are not criticized for getting wrong answers to the solving of their problems.

SUMMARY CONCEPTS:

- Our math classes need a new curriculum in the classroom (i.e., online or face-to-face) because of the challenges students have in understanding and succeeding in their math classes

- Contextualized math curriculum provides students with an engaging class environment whether studying math online or in a face-to-face setting

- Real world math problems assist students in linking mathematical concepts in the classroom, which is important for students to build and retain mathematical knowledge

Chapter Eight

Recommendations and Applications

Recommendations for Practical Applications

One of the first recommendations for college math faculty is to include real world math problems in their lesson plans and curriculum for both developmental (i.e., pre-college math) and college level mathematics (Elcin & Sezer, 2014). These real world problems should be problems that students can relate to in the face-to-face or online math class. For instance, students would be very interested at the college level in getting a new cell phone plan, saving for a used car, or getting a new laptop.

Secondly, college math faculty have an array of different teaching approaches to incorporate real world problem-solving in the math classroom such as through problem-based learning and the use of projects in a math course or using technology (Fatade et al., 2014; Redmond et al., 2011; Wake, 2014).

Thirdly, providing a safe environment for students to experiment, compute, and try different approaches to solving mathematical real world problems will enhance the learning of mathematics (Deed et al., 2012; Marz & Kelchtermans, 2013). Finally, providing a process to facilitate learning for students in the online and face-to-face math classroom will provide students the ability to break the problems down into mini-problems, which will facilitate the learning process (Strimel, 2014).

Recommendations for Future Research

There are a number of areas that can be researched based on the current analysis and understanding of college contextualized mathematics curriculum. Previous research has shown masterfully the effectiveness of contextualized mathematics curriculum both to engage students and assist them in problem solving (Bottge & Cho, 2013; Perin, 2011; Showalter et al., 2013). Hence, future research could further delve into specific types of curricula using contextualized learning such as problem-based learning in mathematics or project-based learning as a comparison of the effectiveness of different types of math curriculum. As another approach to future research, researchers could do a quasi-experimental study on contextualized mathematics classes as a comparison of the effectiveness of the curriculum compared to traditional lecture-based mathematics courses without contextualization. Researchers have provided some evidence with quasi-experimental studies on anchored instruction and modeling-based math classes (Doerr et al., 2014; Elcin & Sezer, 2014; Zydney et al., 2014). Finally, future research could entail a further pedagogical perspective on improving the teaching of college mathematics with contextualized math curriculum.

Summary Concepts and Recommendations:

- Math faculty can include real world math problems in their lesson plans and math curriculum for students at different math levels. Real-world problems should be something students can relate to whether it is renting an apartment, buying a used car, or saving for an upgrade on a new cell phone or new laptop

- Math instructors need different teaching approaches to incorporate real world problem-solving in the math classroom such as through problem-based learning and the use of projects in a math course or through the use of technology such as when using Excel sheets for analysis

- Math teachers need to provide a safe environment for students to experiment, compute, and try different approaches to solving mathematical real-world problems and enhance the learning of mathematics. This can be accomplished by supporting students in the process of solving the math problems rather than focus solely on right or wrong answers

- Math instructors need to facilitate learning with students in the math classroom whether face-to-face or online by providing resources, videos, clear instructions, and math examples

- With the course systems in use today, instructors can implement the use of discussion forums to connect math concepts to real life through questions or assignments in discussion

- Math teachers can provide students with data, instructional videos, and project assignments to connect the math, data, use of concepts to real-life problems such as researching the best cell phone plan to choose

REFERENCES

Adediwura, A. A. (2012). Effect of peer and self-assessment on male and female students' self-efficacy and self-autonomy in the learning of mathematics. *Gender & Behaviour, 10*(1), 4492-4508. Retrieved from http://www.ajol.info/index.php/gab

Albritton, F., Gallard, A. J., & Morgan, M. W. (2010). A comprehensive cost/benefit model: Developmental student success impact. *Journal of Developmental Education, 34*, 10-25. Retrieved from http://ncde.appstate.edu/publications/journal-developmental-education-jde

Al-Huneidi, A. M., & Schreurs, J. (2012). Constructivism-based blended learning in higher education. *International Journal of Emerging Technologies in Learning, 7*(1), 4-9. doi:10.3991/ijet.v7i1.1792

Araujo, Z., Jacobson, E., Singletary, L., Wilson, P., Lowe, L., & Marshall, A. M. (2013). Teachers' conceptions of integrated mathematics curricula. *School Science & Mathematics, 113*(6), 285-296. Retrieved from http://onlinelibrary.wiley.com/journal/10.1111/%28ISSN%291949-8594

Asera, R. (2011). Reflections on developmental mathematics-building new pathways. *Journal of Developmental Education, 34*(3), 28-31. Retrieved from http://ncde.appstate.edu/publications/journal-developmental-education-jde

Ashby, J., Sadera, W., & McNary, S. (2011). Comparing student success between developmental math courses offered online, blended, and face-to-face. *Journal of Interactive Online Learning, 10*(3), 128-140. Retrieved from http://www.ncolr.org/

Baird, K. (2011). Assessing why some students learn math in high school: How useful are student-level test results? *Educational Policy, 25*(5), 784-809. doi:10.1177/0895904810386595

Bandura, A. (1997). *Self-efficacy: The exercise of control.* New York, NY: W.H. Freeman and Company.

Bailey, T., Jeong, D. W., & Cho, S. (2010). Referral, enrollment, and completion in developmental education sequences in community colleges. *Economics of Education Review, 29*(2), 255-270. doi:10.1016/j.econedurev.2009.09.002

Bahr, P. (2012). Deconstructing remediation in community colleges: Exploring associations between course-taking patterns, course outcomes, and attrition from the remedial math and remedial writing sequences. *Research in Higher Education, 53*(6), 661-693. doi:10.1007/s11162-011-9243-2

Battey, D. (2013). 'Good' mathematics teaching for students of color and those in poverty: The importance of relational interactions within instruction. *Educational Studies in Mathematics, 82*(1), 125-144. doi:10.1007/s10649-012-9412-z

Bellamy, J. S., & Mativo, J. M. (2010). A different angle for teaching math. *Technology Teacher, 69*(7), 26-28. Retrieved from http://www.questia.com/library/p5246/the-technology-teacher

Bonham, B. S., & Boylan, H. R. (2011). Developmental mathematics: Challenges, promising practices, and recent initiatives. *Journal of Developmental Education, 36*(2), 14-21. Retrieved from http://ncde.appstate.edu/publications/journal-developmental-education-jde

Bottge, B. A., & Cho, S. J. (2013). Effects of enhanced anchored instruction on skills aligned to Common Core math standards. *Learning Disabilities: A Multidisciplinary Journal, 19*(2), 73-83. Retrieved from http://ldaamerica.org/learning-disabilities-a-multidisciplinary-journal/

Bottge, B. A., Grant, T. S., Rueda, E., & Stephens, A. C. (2010). Advancing the math skills of middle school students in technology education classrooms. *NASSP Bulletin, 94*, 81-106. doi:10.1177/0192636510379902

Bottge, B. A., Ma, X., Gassaway, L., Butler, M., & Toland, M. D. (2014). Detecting and correcting fractions computation error patterns. *Exceptional Children, 80*, 237-255. doi:10.1177/001440291408000207

Briley, J. S. (2012). The relationships among mathematics teaching efficacy, mathematics self-efficacy, and mathematical beliefs for elementary pre-service teachers. *Issues in the Undergraduate Mathematics Preparation of School Teachers, 5*, 1-13. Retrieved from http://www.k-12prep.math.ttu.edu/journal/info-contrib.shtml

Bruner, J. (1977). *The process of education: A landmark in educational history.* Cambridge, MA: Harvard University Press.

Budinski, N., & Takaci, D. (2013). Using Computers and context in the modeling-based teaching of logarithms. *Computers in the Schools, 30*(1), 30-47. doi:10.1080/07380569.2013.764275

Burn, H. E., Baer, E. D., & Wenner, J. M. (2013). Embedded mathematics remediation using the math you need, when you need it: A 21st-century solution to an age-old problem. *About Campus, 18*(5), 22-25. doi:10.1002/abc.21134

Burrows, A., Wickizer, G., Meyer, H., & Borowczak, M. (2013). Enhancing pedagogy with context and partnerships: Science in hand. *Problems of Education in the 21st Century, 54*, 7-13. Retrieved from http://www.jbse.webinfo.lt/Problems_of_Education.htm

Busadee, N., & Laosinchai, P. (2013). Authentic problems in high school probability lesson: Putting research into practice. *Procedia - Social and Behavioral Sciences, 93*, 2043-2047. doi:10.1016/j.sbspro.2013.10.162

Capar, G., & Tarim, K. (2015). Efficacy of the cooperative learning method on mathematics achievement and attitude: A meta-analysis research. *Educational Sciences: Theory & Practice, 15*(2), 553-559. doi:10.12738/estp.2015.2.2098

Castillo-Garsow, C. W. (2014). Mathematical modeling and the nature of problem-solving. *Constructivist Foundations, 9*(3), 373-375. Retrieved from http://www.univie.ac.at/constructivism/journal/

Chang, H. C., Huang, T. H., & Liu, Y. C. (2012). Learning achievement in solving word-based mathematical questions through a computer-assisted learning system. *Educational Technology & Society, 15*(1), 248-259. Retrieved from http://www.ifets.info/index.php

Chun-Yi, L., & Ming-Jang, C. (2015). Effects of worked examples using manipulatives on fifth graders' learning performance and attitude toward mathematics. *Journal of Educational Technology & Society, 18*(1), 264-275. Retrieved from http://www.ifets.info/

Clark, K. R. (2015). The effects of the flipped model of instruction on student engagement and performance in the secondary mathematics classroom. *Journal of Educators Online, 12*(1), 91-115. Retrieved from http://www.thejeo.com

College Board. (2015). Accuplacer math placement tests. Retrieved from https://accuplacer.collegeboard.org/students

Crawford, C., & Persaud, C. (2013). Community colleges online. *Journal of College Teaching & Learning (Online), 10*(1), 75. Retrieved from http://www.cluteinstitute.com/journals/journal-of-college-teaching-learning-tlc/

Deed, C., Pridham, B., Prain, V., & Graham, R. (2012). Drawn into mathematics: Applying student ideas about learning. *International Journal of Pedagogies & Learning, 7*(1), 99-108. doi:10.5172/ijpl.2012.1849

Dennis, J., & O'Hair, M. J. (2010). Overcoming obstacles in using authentic instruction: A comparative case study of high school math & science teachers. *American Secondary Education, 38*(2),

4-22. Retrieved from http://www.ashland.edu/coe/about-college/american-secondary-education-journal

Dewey, J. (2011). *How we think*. Cedar Lake, MI: ReadaClassic.com.

Diaz, C. R. (2010). Transitions in developmental education: An interview with Rosemary Karr. *Journal of Developmental Education, 34*(1), 20-25. Retrieved from http://ncde.appstate.edu/publications/journal-developmental-education-jde

Doerr, H. M., Ärlebäck, J. B., & Costello Staniec, A. (2014). Design and effectiveness of modeling-based mathematics in a summer bridge program. *Journal of Engineering Education, 103*(1), 92-114. doi:10.1002/jee.20037

Duhon, G. J., House, S., Hastings, K., Poncy, B., & Solomon, B. (2015). Adding immediate feedback to explicit timing: An option for enhancing treatment intensity to improve mathematics fluency. *Journal of Behavioral Education, 24*(1), 74-87. doi:10.1007/s10864-014-9203-y

Elcin, M., & Sezer, B. (2014). An exploratory comparison of traditional classroom instruction and anchored instruction with secondary school students: Turkish experience. *Eurasia Journal of Mathematics, Science & Technology Education, 10*(6), 523-530. doi:10.12973/eurasia.2014.1171a

Erbas, A. K., Ince, M., & Kaya, S. (2015). Learning mathematics with interactive whiteboards and computer-based graphing utility. *Journal of Educational Technology & Society, 18*(2), 299-312. Retrieved from http://www.ifets.info

Erbas, A. K., Kertil, M., Cetinkaya, B., Cakiroglu, E., Alacadi, C., & Bas, S. (2014). Mathematical modeling in mathematics education: Basic concepts and approaches. *Educational Sciences: Theory & Practice, 14*(5), 1621-1627. doi:10.12738/estp.2014.4.2039

Erlich, R. J., & Russ-Eft, D. (2011). Applying social cognitive theory to academic advising to assess student learning outcomes. *NACADA Journal, 31*(2), 5-15. doi:10.12930/0271-9517-31.2.5

Fan, L., & Bokhove, C. (2014). Rethinking the role of algorithms in school mathematics: A conceptual model with focus on cognitive development. *ZDM, 46*(3), 481-492. doi:10.1007/s11858-014-0590-2

Fast, G. R., & Hankes, J. E. (2010). Intentional integration of mathematics content instruction with constructivist pedagogy in elementary mathematics education. *School Science and Mathematics, 110*(7), 330-340. doi:10.1111/j.1949-8594.2010.00043.x

Fatade, A. O., Arigbabu, A. A., Mogari, D., & Awofala, A. A. (2014). Investigating senior secondary school students' beliefs about further mathematics in a problem-based learning context. *Bulgarian Journal of Science & Education Policy, 8*(1), 5-46. Retrieved from http://bjsep.org/index.php

Gardner, H. (2006). *Multiple intelligences: New horizons*. New York, NY: Basic Books.

Gardner, H. (2011). *Frames of mind: The theory of multiple intelligences*. New York, NY: Basic Books.

George, M. (2010). Ethics and motivation in remedial mathematics education. *Community College Review, 38*(1), 82-92. doi:10.1177/0091552110373385

Goeller, L. (2013). Developmental mathematics: Students' perceptions of the placement process. *Research & Teaching in Developmental Education, 30*(1), 22-34. Retrieved from http://www.nyclsa.org/journal.html

Goodman, R. (2010). Problem-based learning: Merging of economics and mathematics. *Journal of Economics & Finance, 34*(4), 477-483. doi:10.1007/s12197-010-9154-7

Grady, M., Watkins, S., & Montalvo, G. (2012). The effect of constructivist mathematics on achievement in rural schools. *Rural Educator, 33*(3), 38-47. Retrieved from http://www.ruraleducator.net/

Gucler, B. (2014). The role of symbols in mathematical communication: The case of the limit notation. *Research in Mathematics Education, 16*(3), 251-268. doi:10.1080/14794802.2014.919872

Gunbas, N. (2015). Students' mathematics word problem-solving achievement in a computer-based story. *Journal of Computer Assisted Learning, 31*(1), 78-95. doi:10.1111/jcal.12067

Hanim, S. Z., Mohd, F. E., & Zainol, I. I. (2012). Student-centred learning in mathematics - constructivism in the classroom. *Journal of International Education Research, 8*(4), 319-328. Retrieved from www.cluteinstitute.com/journals/journal-of-international-education-research-jier/

Harvey, R., & Averill, R. (2012). A lesson based on the use of contexts: An example of effective practice in secondary school mathematics. *Mathematics Teacher Education & Development, 14*(1), 41-59. Retrieved from http://www.merga.net.au/publications/mted.php

Hasan, A., & Fraser, B. J. (2015). Effectiveness of teaching strategies for engaging adults who experienced childhood difficulties in learning mathematics. *Learning Environments Research, 18*(1), 1-13. doi:10.1007/s10984-013-9154-6

Hennessey, M. N., Higley, K., & Chesnut, S. R. (2012). Persuasive pedagogy: A new paradigm for mathematics education. *Educational Psychology Review, 24*(2), 187-204. doi:10.1007/s10648-011-9190-7

Hern, K. (2012). Acceleration across California: Shorter pathways in developmental English and math. *Change, 44*(3), 60-68. doi:10.1080/00091383.2012.672917

Howard, L., & Whitaker, M. (2011). Unsuccessful and successful mathematics learning: Developmental students' perceptions. *Journal of Developmental Education, 35*(2), 2-16. Retrieved from http://ncde.appstate.edu/publications/journal-developmental-education-jde

Howell, J. S. (2011). What influences students' need for remediation in college? Evidence from California. *Journal of Higher Education, 82*(3), 292-318. doi:10.1353/jhe.2011.0014

Huang, Y., Huang, S., & Wu, T. (2014). Embedding diagnostic mechanisms in a digital game for learning mathematics.

Educational Technology, Research and Development, 62(2), 187-207. doi:10.1007/s11423-013-9315-4

Hudesman, J., Crosby, S., Ziehmke, N., Everson, H., Isaac, S., Flugman, B., Moylan, A. (2014). Using formative assessment and self-regulated learning to help developmental mathematics students achieve: A multi-campus program. *Journal on Excellence in College Teaching, 25*(2), 107-121. Retrieved from http://celt.muohio.edu/ject

Joet, G., Usher, E. L., & Bressoux, P. (2011). Sources of self-efficacy: An investigation of elementary school students in France. *Journal of Educational Psychology, 103*(3), 649-663. doi:10.1037/a0024048

Jones, S. J., & Long, V. M. (2013). Learning equity between online and on-site mathematics courses. *Journal of Online Learning & Teaching, 9*(1), 1-12. Retrieved from http://jolt.merlot.org/

Jorgensen, M. E. (2010). Questions for practice: Reflecting on developmental mathematics using 19th-century voices. *Journal of Developmental Education, 34*(1), 26-28. Retrieved from http://ncde.appstate.edu/publications/journal-developmental-education-jde

Khiat, H. (2010). A grounded theory approach: Conceptions of understanding in engineering mathematics learning. *Qualitative Report, 15*(6), 1459-1488. Retrieved from http://www.nova.edu/ssss/QR/about.html

Kim, M. K., & Cho, M. K. (2015). Design and implementation of integrated instruction of mathematics and science in Korea.

Eurasia Journal of Mathematics, Science & Technology Education, 11(1), 3-14. doi:10.12973/eurasia.2015.1301a

Kinnari-Korpela, H. (2015). Using short video lectures to enhance mathematics learning - experiences on differential and integral calculus course for engineering students. *Informatics in Education, 14*(1), 67-81. doi:10.15388/infedu.2015.05

Krummheuer, G. (2013). The relationship between diagrammatic argumentation and narrative argumentation in the context of the development of mathematical thinking in the early years. *Educational Studies in Mathematics, 84*(2), 249-265. doi:10.1007/s10649-013-9471-9

Laah-On, S., Intaros, P., & Sangaroon, K. (2013). Key universal activities of mathematical learning in problem solving mathematics classroom. *Creative Education, 4*(11), 700-704. doi:10.4236/ce.2013.411099

Lavidas, K., Komis, V., & Gialamas, V. (2013). Spreadsheets as cognitive tools: A study of the impact of spreadsheets on problem-solving of math story problems. *Education and Information Technologies, 18*(1), 113-129. doi:dx.doi.org/10.1007/s10639-011-9174-8

Levpuscek, M. P., Zupancic, M., & Socan, G. (2013). Predicting achievement in mathematics in adolescent students: The role of individual and social factors. *Journal of Early Adolescence, 33*(4), 523-551. doi:10.1177/0272431612450949

Lundin, S. (2012). Hating school, loving mathematics: On the ideological function of critique and reform in mathematics education.

Educational Studies in Mathematics, 80(1), 73-85. doi:10.1007/ s10649-011-9366-6

Lynott, F. J., Kracl, C. L., Knoell, C. M., & Harshbarger, D. (2013). Using the shared integration approach: A more "authentic approach" to middle school interdisciplinary lessons in health, mathematics, science, and literature. *Strategies: A Journal for Physical and Sport Educators, 26*(3), 13-18. doi:10.1080/08924562.2013.779862

Marz, V., & Kelchtermans, G. (2013). Sense-making and structure in teachers' reception of educational reform. A case study on statistics in the mathematics curriculum. *Teaching and Teacher Education, 29*, 13-24. doi:10.1016/j.tate.2012.08.004

Mativo, J. M., Hill, R. B., & Godfrey, P. W. (2013). Effects of human factors in engineering and design for teaching mathematics: A comparison study of online and face-to-face at a technical college. *Journal of STEM Education: Innovations & Research, 14*(4), 36-44. Retrieved from http://ojs.jstem.org/index.php?journal=JSTEM

Merseth, K. K. (2011). Update: Report on innovations in developmental mathematics—moving mathematical graveyards. *Journal of Developmental Education, 34*(3), 32-39. Retrieved from http://ncde.appstate.edu/publications/journal-developmental-education-jde

Mhakure, D., & Mokoena, M. A. (2011). A comparative study of the FET phase mathematical literacy and mathematics curriculum. *Online Submission, 3*, 309-323. Retrieved from http://www.davidpublishing.org/journals_info.asp?jId=506

Miller, M. D. (2011). What college teachers should know about memory: A perspective from cognitive psychology. *College Teaching, 59*(3), 117-122. doi:10.1080/87567555.2011.580636

Milner, A. R., Templin, M. A., & Czerniak, C. M. (2011). Elementary science students' motivation and learning strategy use: Constructivist classroom contextual factors in a life science laboratory and a traditional classroom. *Journal of Science Teacher Education, 22*(2), 151-170. doi:10.1007/s10972-010-9200-5

Ngu, B. H., & Yeung, A. S. (2013). Algebra word problem-solving approaches in a chemistry context: Equation worked examples versus text editing. *Journal of Mathematical Behavior, 32,* 197-208. doi:10.1016/j.jmathb.2013.02.003

Nijlen, D. V., & Janssen, R. (2015). Examinee non-effort on contextualized and non-contextualized mathematics items in large-scale assessments. *Applied Measurement in Education, 28*(1), 68-84. doi:10.1080/08957347.2014.973559

O'Brien, T. C., Wallach, C., & Mash-Duncan, C. (2011). Problem-based learning in mathematics. *Montana Mathematics Enthusiast, 8*(1/2), 147-159. Retrieved from http://www.math.umt.edu/tmme/

O'Shea, J., & Leavy, A. (2013). Teaching mathematical problem-solving from an emergent constructivist perspective: The experiences of Irish primary teachers. *Journal of Mathematics Teacher Education, 16*(4), 293-318. doi:10.1007/s10857-013-9235-6

Perin, D. (2011). Facilitating student learning through contextualization: A review of evidence. *Community College Review, 39*(3), 268-295. doi:10.1177/0091552111416227

Petty, T., & Farinde, A. A. (2013). Investigating student engagement in an online mathematics course through windows into teaching and learning. *Journal of Online Learning & Teaching, 9*(2), 261-270. Retrieved from http://jolt.merlot.org/

Piaget, J. (2001). *The psychology of intelligence.* New York, NY: Routledge.

Prado, M. M., & Gravoso, R. S. (2011). Improving high school students' statistical reasoning skills: A case of applying anchored instruction. *Asia-Pacific Education Researcher, 20*(1), 61-72. Retrieved from http://link.springer.com/journal/40299

Puri, K., Cornick, J., & Guy, M. (2014). An analysis of the impact of course elimination via contextualization in developmental mathematics. *MathAMATYC Educator, 5*(2), 4-10. Retrieved from http://www.amatyc.org/?page=MathAMATYCEducator

Redmond, A., Thomas, J., High, K., Scott, M., Jordan, P., & Dockers, J. (2011). Enriching science and math through engineering. *School Science and Mathematics, 111*(8), 399-408. doi:10.1111/j.1949-8594.2011.00105.x

Reisel, J. R., Jablonski, M. R., Munson, E., & Hosseini, H. (2014). Peer-led team learning in mathematics courses for freshmen engineering and computer science students. *Journal of STEM Education: Innovations & Research, 15*(2), 7-15. Retrieved from http://www.jstem.org/

Roykenes, K. (2015). "My math and me": Nursing students' previous experiences in learning mathematics. *Nurse Education in Practice,* 2015, 1-7. doi:10.1016/j.nepr.2015.05.009

Sahin, I. (2010). Curriculum assessment: Constructivist primary mathematics curriculum in Turkey. *International Journal of Science & Mathematics Education, 8*(1), 51-72. doi:10.1007/s10763-009-9162-2

Schwartz, C. (2012). Counting to 20: Online implementation of a face-to-face, elementary mathematics methods problem-solving activity. *Techtrends: Linking Research & Practice to Improve Learning, 56*(1), 34-39. doi:10.1007/s11528-011-0551-3

Schwols, A., & Miller, K. B. (2012). Identifying mathematics content and integrating it into science instruction. *Science Scope, 36*(1), 48-52. Retrieved from https://www.questia.com/library/p2623/science-scope

Sheldon, C. Q., & Durdella, N. R. (2010). Success rates for students taking compressed and regular length developmental courses in the community college. *Community College Journal of Research & Practice, 34*(1), 39-54. doi:10.1080/10668920903385806

Shiu, E. (2013). Improving learning in statistics – A pilot trial study to achieve the triple effects of depth, breadth and integration. *International Journal of Management Education, 11*, 12-24. doi:10.1016/j.ijme.2012.11.001

Showalter, D. A., Wollett, C., & Reynolds, S. (2014). Teaching a high-level contextualized mathematics curriculum to adult basic learners. *Journal of Research & Practice for Adult Literacy, Secondary & Basic Education, 3*(2), 21-34. Retrieved from http://www.coabe.org/html/abeljournal.html

Sidney, P. G., & Alibali, M. W. (2015). Making connections in math: Activating a prior knowledge analogue matters for learning. *Journal of Cognition and Development, 16*(1), 160-185. doi:10.10 80/15248372.2013.792091

Spradlin, K., & Ackerman, B. (2010). The effectiveness of computer assisted instruction in developmental mathematics. *Journal of Developmental Education, 34*(2), 12-42. Retrieved from http://ncde.appstate.edu/publications/journal-developmental-education-jde

Strimel, G. (2014). Authentic education. *Technology & Engineering Teacher, 73*(7), 8-18. Retrieved from http://www.iteea.org/Publications/ttt.htm

Tammaro, R., & Solco, S. (2013). From knowledge to action to become competent people: From traditional to authentic assessment. *Knowledge Cultures, 1*(6), 147-163. Retrieved from http://www.addletonacademicpublishers.com/knowledge-cultures

The International GeoGebra Institute. (2015). GeoGebra. Retrieved from https://www.geogebra.org/

Toews, C. (2012). Mathematical modeling in the undergraduate curriculum. *Primus: Problems, Resources, and Issues in Mathematics Undergraduate Studies, 22*(7), 545-563. doi:10.1080/10511970.2 011.648003

Tosmur-Bayazit, N., & Ubuz, B. (2013). Practicing engineers' perspective on mathematics and mathematics education in college. *Journal of STEM Education: Innovations & Research, 14*(3), 34-40. Retrieved from http://ojs.jstem.org/index.php?journal=JSTEM

Trinter, C. P., Moon, T. R., & Brighton, C. M. (2015). Characteristics of students' mathematical promise when engaging with problem-based learning units in primary classrooms. *Journal of Advanced Academics, 26*(1), 24-58. doi:10.1177/1932202X14562394

Ulrich, C., Tillema, E. S., Hackenberg, A. J., & Norton, A. (2014). Constructivist model building: Empirical examples from mathematics education. *Constructivist Foundations, 9*(3), 328-339. Retrieved from http://www.univie.ac.at/constructivism/journal/

Valenzuela, H. (2012). An integration of math with auto technician courses. *MathAMATYC Educator, 3*(2), 53-56. Retrieved from http://www.amatyc.org/?page=MathAMATYCEducator

Valenzuela, H. (2014). Preparing English as a second language students for college level math. *MathAMATYC Educator, 5*(2), 12-15. Retrieved from http://www.amatyc.org/?page=MathAMATYCEducator

Vanags, T., Pammer, K., & Brinker, J. (2013). Process-oriented guided-inquiry learning improves long-term retention of information. *Advances in Physiology Education, 37*(3), 233-241. doi:10.1152/advan.00104.2012

VanOra, J. (2012). The experience of community college for developmental students: Challenges and motivations. *The Community College Enterprise, 18*(1), 22-36. Retrieved from https://www.schoolcraft.edu/a-z-index/community-college-enterprise#.U4XxwvlkSSo

Varsavsky, C. (2010). Chances of success in and engagement with mathematics for students who enter university with a weak mathematics background. *International Journal of Mathematical Education in Science & Technology, 41*(8), 1037-1049. doi:10.1080/0020739X.2010.493238

Vasquez Mireles, S. (2010). Theory to practice developmental mathematics program: A model for change. *Journal of College Reading and Learning, 40*(2), 81-90. Retrieved from http://www.crla.net/journal.htm

Venezia, A., & Jaeger, L. (2013). Transitions from high school to college. *Future of Children, 23*(1), 117-136. doi:10.1353/foc.2013.0004

Vygotsky, L. S. (1978). *Mind in society: The development of higher psychological processes.* Cambridge, MA: Harvard University Press.

Wake, G. (2014). Making sense of and with mathematics: The interface between academic mathematics and mathematics in practice. *Educational Studies in Mathematics, 86*(2), 271-290. doi:10.1007/s10649-014-9540-8

Waycaster, P. (2011). Tracking developmental students into their first college level mathematics course. *Inquiry, 16*(1), 53-66. Retrieved from http://www.vccaedu.org/inquiry/

Weigmann, K. (2013). Educating the brain: The growing knowledge about how our brain works can inform educational programmes and approaches, in particular, for children with learning problems. *EMBO Reports, 14*(2), 136-139. doi:10.1038/embor.2012.213

Weiss, M., Herbst, P., & Chen, C. (2009). Teachers' perspectives on "authentic mathematics" and the two-column proof form. *Educational Studies in Mathematics, 70*(3), 275-293. doi:10.1007/s10649-008-9144-2

Wenner, J. M., Burn, H. E., & Baer, E. M. (2011). The math you need, when you need it: Online modules that remediate mathematical skills in introductory geoscience courses. *Journal of College Science Teaching, 41*(1), 16-24. Retrieved from http://www.nsta.org/college/

Wilson, F. C. (2011). A make it real approach. *MathAMATYC Educator, 3*(1), 52-53. Retrieved from http://www.amatyc.org/?page=MathAMATYCEducator

Wood, L. N. (2010). Graduate capabilities: Putting mathematics into context. *International Journal of Mathematical Education in Science & Technology, 41*(2), 189-198. doi:10.1080/00207390903388607

Young, R., Hodge, A., Edwards, M., & Leising, J. (2012). Learning mathematics in high school courses beyond mathematics: Combating the need for post-secondary remediation in mathematics. *Career & Technical Education Research, 37*(1), 21-33. doi:10.5328/cter37.1.21

Yu-Liang, C. (2012). A study of fifth graders' mathematics self-efficacy and mathematical achievement. *Asia-Pacific Education Researcher, 21*(3), 519-525. Retrieved from http://www.dlsu.edu.ph/research/journals/taper/

Zahner, W., Velazquez, G., Moschkovich, J., Vahey, P., & Lara-Meloy, T. (2012). Mathematics teaching practices with technology that support conceptual understanding for Latino/a students. *Journal of Mathematical Behavior, 31*(4), 431-446. doi:10.1016/j.jmathb.2012.06.002

Zain, S. S., Rasidi, F. M., & Abidin, I. Z. (2012). Student-centered learning in mathematics - constructivism in the classroom. *Journal of International Education Research, 8*(4), 319-328. Retrieved from http://www.cluteinstitute.com/journals/journal-of-international-education-research-jier/

Zydney, J. M., Bathke, A., & Hasselbring, T. S. (2014). Finding the optimal guidance for enhancing anchored instruction. *Interactive Learning Environments, 22*(5), 668-683. doi:10.1080/10494820.2012.745436

CPSIA information can be obtained
at www.ICGtesting.com
Printed in the USA
BVHW092040251021
619818BV00009B/254